KU-252-919

THE HUMAN ZOO

By the same author

THE TEN-SPINED STICKLEBACK
THE BIOLOGY OF ART
THE MAMMALS: A GUIDE TO THE LIVING SPECIES
PRIMATE ETHOLOGY (EDITOR)
THE NAKED APE

With Ramona Morris

MEN AND SNAKES
MEN AND APES
MEN AND PANDAS

For Children

THE STORY OF CONGO
MONKEYS AND APES
THE BIG CATS
ZOOTIME

DESMOND MORRIS

THE HUMAN ZOO

WORLD BOOKS · LONDON

This edition published 1971
by World Books by arrangement
with Jonathan Cape Ltd

*Printed in Great Britain by Richard Clay (The Chaucer Press), Ltd.,
Bungay, Suffolk*

CONTENTS

ACKNOWLEDGMENTS

As with its predecessor, *The Naked Ape*, this book is intended for a general audience and authorities have therefore not been quoted in the text. However, many original books and papers have been referred to during the assembly of this volume and it would be wrong to present it without acknowledging their valuable assistance. On pages 246–8 I have included a chapter-by-chapter appendix relating the topics discussed to the major authorities concerned. This appendix can be used to trace the detailed references given in the selected bibliography.

I would also like to express my debt and my gratitude to the many colleagues and friends who have helped me, in discussions, correspondence and many other ways. Their contributions have varied. In some instances, they have been of direct assistance in connection with a specific point in the present text, but in other cases they have been stimulating in a more indirect way, often over a period of years, influencing my general thinking and helping me to clarify my views. With a subject as broad as *The Human Zoo*, it is impossible to name them all, but they include, in particular, the following: Dr Anthony Ambrose, Mr Robert Ardrey, Mr David Attenborough, Mr Kenneth Bayes, Professor Misha Black, Dr David Blest, Dr N. G. Blurton-Jones, Mr James Bomford, Dr John Bowlby, Mr Richard Carrington, Sir Hugh Casson, Dr Michael Chance, Dr Richard Cross, Dr Christopher Evans, Professor Robin Fox, Professor J. H. Fremlin, Mr Oliver Graham-Jones, Dr Fae Hall, Professor Harry Harlow, Mrs Mary Haynes, Professor Heini Hediger, Professor Robert Hinde, Dr Jan van Hooff, Dr Francis

Huxley, Sir Julian Huxley, Professor Janey Ironside, Miss Devra Kleiman, Dr Adriaan Kortlandt, Baroness Jane van Lawick-Goodall, Dr Paul Leyhausen, Mrs Caroline Loizos, Professor Konrad Lorenz, Dr Malcolm Lyall-Watson, Dr Gilbert Manley, Dr Isaac Marks, Mr Tom Maschler, Dr L. Harrison Matthews, Lady Medway, Mrs Ramona Morris, Dr Martin Moynihan, Dr John Napier, Mrs Caroline Nicolson, Mr Philip Oakes, Dr Kenneth Oakley, Mr Victor Pasmore, Sir Roland Penrose, Sir Herbert Read, Dr Frances Reynolds, Dr Vernon Reynolds, Mrs Claire Russell, Dr W. M. S. Russell, Professor Arthur Smailes, Mr Peter Shepheard, Dr John Sparks, Dr Anthony Storr, Mr Frank Taylor, Dr Lionel Tiger, Professor Niko Tinbergen, Dr Nevil Tronchin-James, Mr Ronald Webster, Dr Wolfgang Wickler, Miss Pat Williams, Dr G. M. Woddis and Professor John Yudkin.

I hasten to add that the inclusion of a name in this list does not imply that the person concerned necessarily agrees with my views as expressed in this book.

INTRODUCTION

WHEN the pressures of modern living become heavy, the harassed city-dweller often refers to his teeming world as a concrete jungle. This is a colourful way of describing the pattern of life in a dense urban community, but it is also grossly inaccurate, as anyone who studied a real jungle will confirm.

Under normal conditions, in their natural habitats, wild animals do not mutilate themselves, masturbate, attack their offspring, develop stomach ulcers, become fetishists, suffer from obesity, form homosexual pair-bonds or commit murder. Among human city-dwellers, needless to say, all of these things occur. Does this, then, reveal a basic difference between the human species and other animals? At first glance it seems to do so. But this is deceptive. Other animals do behave in these ways under certain circumstances, namely when they are confined in the unnatural conditions of captivity. The zoo animal in a cage exhibits all these abnormalities that we know so well from our human companions. Clearly, then, the city is not a concrete jungle, it is a human zoo.

The comparison we must make is not between the city-dweller and the captive animal, but between the city-dweller and the captive animal. The modern human animal is no longer living in conditions natural for his species. Trapped, not by a zoo collector, but by his own brainy brilliance, he has set himself up in a huge, restless menagerie where he is in constant danger of cracking under the strain.

Despite the pressures, however, the benefits are great. The zoo world, like a gigantic parent, protects its inmates:

food, drink, shelter, hygiene and medical care are provided; the basic problems of survival are reduced to a minimum. There is time to spare. How this time is used in a non-human zoo varies, of course, from species to species. Some animals quietly relax and doze in the sun; others find prolonged inactivity increasingly difficult to accept. If you are an inmate of a human zoo, you inevitably belong to this second category. Having an essentially exploratory, inventive brain, you will not be able to relax for very long. You will be driven on and on to more and more elaborate activities. You will investigate, organize and create and, in the end, you will have plunged yourself deeper still into an even more captive zoo world. With each new complexity, you will find yourself one step farther away from your natural tribal state, the state in which your ancestors existed for a million years.

The story of modern man is the story of his struggle to deal with the consequences of this difficult advance. The picture is confused and confusing, partly because of its very complexity and partly because we are involved in it in a dual role, being, at the same time, both spectators and participants. Perhaps it will become clearer if we view it from the zoologist's standpoint, and this is what I shall attempt to do in the pages that follow. In most cases I have deliberately selected examples which will be familiar to Western readers. This does not mean, however, that I intended my conclusions to relate only to Western cultures. On the contrary, there is every indication that the underlying principles apply equally to city-dwellers throughout the world.

If I seem to be saying 'Go back, you are heading for disaster,' let me assure you that I am not. We have, in our relentless social progress, gloriously unleashed our powerful inventive, exploratory urges. They are a basic part of our biological inheritance. There is nothing artificial or unnatural about them. They provide us with our great strength as well as our great weaknesses. What I am trying to show is the increasing price we have to pay for indulging them and the ingenious ways in which we contrive to meet that price, no matter how steep it becomes. The stakes are rising higher all the time, the game becoming more risky, the casualties more startling, the pace more breathless. But despite the hazards it is the most exciting game the world has ever seen. It is foolish to suggest that anyone should blow a whistle and try to stop it. Nevertheless, there are different ways of playing it, and if we can understand better the true nature of the players it should be possible to make the game even more rewarding, without at the same time becoming more dangerous and, ultimately, disastrous for the whole species.

CHAPTER ONE

TRIBES AND SUPER-TRIBES

IMAGINE a piece of land twenty miles long and twenty miles wide. Picture it wild, inhabited by animals small and large. Now visualize a compact group of sixty human beings camping in the middle of this territory. Try to see yourself sitting there, as a member of this tiny tribe, with the landscape, your landscape, spreading out around you farther than you can see. No one apart from your tribe uses this vast space. It is your exclusive home-range, your tribal hunting ground. Every so often the men in your group set off in pursuit of prey. The women gather fruit and berries. The children play noisily around the camp site, imitating the hunting techniques of their fathers. If the tribe is successful and swells in size, a splinter group will set off to colonize a new territory. Little by little the species will spread.

Imagine a piece of land twenty miles long and twenty miles wide. Picture it civilized, inhabited by machines and buildings. Now visualize a compact group of six million human beings camping in the middle of this territory. See yourself sitting there, with the complexity of the huge city spreading out all around you, farther than you can see.

Now compare these two pictures. In the second scene there are a hundred thousand individuals for every one in the first scene. The space has remained the same. Speaking in evolutionary terms, this dramatic change has been almost instantaneous; it has taken a mere few thousand years to convert scene one into scene two. The human animal appears to have adapted brilliantly to his extraordinary new condition, but he has not had time to change biologically, to evolve into a new, genetically civilized species. This civilizing process has been accomplished entirely by learning and conditioning. Biologically he is still the simple tribal animal depicted in scene one. He lived like that, not for a few centuries, but for a million hard years. During that period he did change biologically. He evolved spectacularly. The pressures of survival were great and they moulded him.

So much has happened in the past few thousand years, the urban years, the crowded years of civilized man, that we find it hard to grasp the idea that this is no more than a minute part of the human story. It is so familiar to us that we vaguely imagine we grew into it gradually and that, as a result, we are biologically fully equipped to deal with all the new social hazards. If we force ourselves to be coolly objective about it, we are bound to admit that this is not so. It is only our incredible plasticity, our ingenious adaptability, that makes it seem so. The simple tribal hunter is doing his best to wear his new trappings lightly and proudly; but they are complex, cumbersome garments and he keeps tripping over them. However, before we examine the way he trips and so frequently loses his balance, we must first see how he contrived to stitch together his fabulous cloak of civilization.

We must begin by lowering the temperature until we are back in the grip of the Ice Age, say, twenty thousand years ago. Our early hunting ancestors had already succeeded in spreading throughout much of the Old World and were soon to trek across from eastern Asia to the New World. To have achieved such a shattering expansion must have meant that their simple hunting way of life was already more than a match for their carnivorous rivals. But this is not so surprising when one stops to think that our Ice Age ancestors' brains were already as big and highly developed as ours are today. Skeletally, there is little to choose between us. The modern man, physically speaking, had already arrived on the scene. In fact, if it were possible, with the aid of a time machine, to take the newborn child of an Ice Age hunter into your home and rear it as your own, it is doubtful if anyone would detect the deception.

In Europe the climate was hostile, but our ancestors fought it well. With the simplest of technologies they were able to slay huge game animals. Happily they have left us a testimony of their hunting skills, not only in the accidental remnants that we can scratch up from the floors of their caves, but also in the staggering murals painted on their walls. The shaggy mammoths, woolly rhinos, bison and reindeer portrayed there leave no doubt as to the nature of the climate. As one emerges from the darkness of the caves today and steps out into the baked countryside, it is difficult to imagine it inhabited by these heavy-furred creatures. The contrast between the temperature then and now comes vividly to mind.

As the last glaciation came towards its end, the ice began to retreat northwards at a rate of fifty yards a year and the cold-country animals moved north with it. Rich forests took the place of the cold tundra landscape. The great Ice Age ended about ten thousand years ago and heralded a new epoch in human development.

The breakthrough was to come at the point where Africa, Asia and Europe meet. There, at the eastern end of the Mediterranean, there was a small change in human feeding behaviour that was to alter the whole course of man's progress. It was trivial enough and simple enough in itself, but its impact was to be enormous. Today we take it for granted: we call it farming.

Previously all human tribes had filled their bellies in one

of two ways: the men had hunted for animal foods and the women had gathered plant foods. The diet was balanced by the sharing of the spoils. Virtually all the active adult members of the tribe were food-getters. There was comparatively little food-storing. They merely went out and collected what they wanted when they wanted it. This was less hazardous than it sounds because, of course, the whole world population of our species was then minute, compared with the massive numbers of today. However, although these early hunter/gatherers were highly successful and spread to cover a large part of the globe, their tribal units remained small and simple. During the hundreds of thousands of years of human evolution, men had become increasingly adapted, both physically and mentally, both structurally and behaviourally, to this hunting way of life. The new step they took, the step to farming and food-production, swept them over an unexpected threshold and threw them so rapidly into an unfamiliar form of social existence, that there was no time for them to evolve new, genetically-controlled qualities to go with it. From now on, their adaptability and behavioural plasticity, their ability to learn and adjust to novel and more complex ways, were going to be tested to the full. Urbanization and the intricacies of town-living were only one more step away.

Luckily the long hunting apprenticeship had developed ingenuity and a mutual-aid system. The human hunters, it is true, were still innately competitive and self-assertive, like their monkey ancestors, but their competitiveness had become forcibly tempered by an increasingly basic urge to co-operate. It had been their only hope of succeeding in their rivalry with the long-established, sharp-clawed, professional killers of the carnivore world, such as the big cats. The human hunters had evolved their co-operativeness alongside their intelligence and their exploratory nature, and the combination had proved effective and deadly. They learned quickly, remembered well and were adept at bringing together separate elements of their past learning to solve new problems. If this quality had been helpful to them in the early days, when on their arduous hunting trips, it was even more essential to them now, nearer to home, as they stood on the threshold of a new and vastly more complex form of social life.

The lands around the eastern end of the Mediterranean were the natural homes of two vital plants: wild wheat and wild barley. Also found in this region were wild goats, wild sheep, wild cattle and wild pigs. The human hunter/gatherers who settled in this area had already domesticated the dog, but it was used primarily as a hunting companion and watchdog, rather than as a direct source of food. True farming began with the cultivation of the two plants, wheat and barley. It was followed soon after by the domestication, first, of goats and sheep, and then, a little later, of cattle and pigs. In all probability, the animals were first attracted by the cultivated crops, came to eat and stayed to be bred and eaten themselves.

It is no accident that the other two regions of the earth which, later on, saw the birth of independent ancient civilizations (southern Asia and central America) were also places where the hunter/gatherers found wild plants suitable for cultivation: rice in Asia and maize in America.

So successful were these Late Stone Age cultivations that, from that day to this, the plants and animals that were domesticated then have remained the major food sources of all large-scale agricultural operations. The great new advance in farming has been mechanical rather than biological. But it was what started out as the mere left-overs of early farming that were to have the truly shattering impact on our species.

Retrospectively it is easy to explain. Before the arrival of farming, everyone who was going to eat had to do their share of food-finding. Virtually the whole tribe was involved. But when the forward-thinking brains that had planned and schemed hunting manoeuvres turned their attention to the problems of organizing the cultivation of crops, the irrigation of land and the breeding of captive animals, they achieved two things. Such was their success that they created, for the first time, not only a constant food supply, but also a regular and reliable food *surplus*. The creation of this surplus was the key that was to unlock the gateway to civilization. At last, the human tribe could support more members than were required to find food. The tribe could not only become bigger, but it could free some of its people for other tasks: not part-time tasks, fitted in around the priority demands of food-finding, but full-time activities that could flourish in their own right. An age of specialization was born.

From these small beginnings grew the first towns.

I have said that it is easy to explain, but all this means is that it is not difficult for us, in looking back, to pick out the vital factor that led on to the next great step in the human story. It does not, of course, mean that it was an easy step to take at the time. It is true that the human hunter/gatherer was a magnificent animal, full of untapped potentialities and capabilities. The fact that we are here today is proof enough. But he had evolved as a tribal hunter, not as a patient, sedentary farmer. It is also true that he had a far-seeing brain, capable of planning a hunt and understanding the seasonal changes in his environment. But to be a successful farmer, he had to stretch his far-seeingness beyond anything he had previously experienced. The tactics of hunting had to become the strategy of farming. This accomplished, he had to push his brain even farther to contend with the added social complexities that were to follow his new-found affluence, as villages grew into towns.

It is important to realize this when talking of an 'urban revolution'. The use of this phrase gives the impression that towns and cities began to spring up all over the place in an overnight rush towards an impressive new social life. But it was not like that. The old ways died hard and slowly. Indeed, in many parts of the world they are still with us today. Numerous contemporary cultures are still operating at virtually Neolithic farming levels, and in certain regions, such as the Kalahari Desert, Northern Australia, and the Arctic, we can still observe Palaeolithic-style communities of hunter/gatherers.

The first urban developments, the first towns and cities, grew, not as a sudden rash on the skin of prehistoric society, but as a few tiny, isolated spots. They appeared at sites in south-west Asia as dramatic exceptions to the general rule. By present-day standards they were very small and the pattern spread slowly, very slowly. Each was based on a highly localized organization, intimately connected with and bound up in the surrounding farmlands.

At first there was little trade or inter-action between one urban centre and the others. This was to be the next great advance, and it took time. The psychological barrier to such a step was obviously the loss of local identity. It was not so much a case of 'the tribe that lost its head', as the human head refusing to lose its tribe. The species had evolved as a tribal animal and the basic characteristic of a tribe is that it operates on a localized, inter-personal basis. To abandon this fundamental social pattern, so typical of the ancient human condition, was going to go against the grain. But it was the grain, in another sense, so efficiently harvested and transported, that was forcing the pace. As agriculture advanced and the urban elite, liberated from the labours of production, began to concentrate their brain-power on other, newer problems, it was inevitable that there would eventually emerge an urban network, a hierarchically organized interconnection between neighbouring towns and cities.

The oldest known town arose at Jericho more than 8,000 years ago, but the first fully urban civilization developed farther east, across the Syrian desert in Sumer. There, between 5,000 and 6,000 years ago, the first empire was born, and the 'pre' was taken out of prehistory with the invention of writing. Inter-city co-ordination developed, leaders became administrators, professions became established, metalwork and transport advanced, beasts of burden (as distinct from food animals) were domesticated and monumental architecture arose.

By our standards the Sumerian cities were small, with populations ranging from 7,000 to no more than 20,000. Nevertheless, our simple tribesman had already come a long way. He had become a citizen, a super-tribesman, and the key difference was that *in a super-tribe he no longer knew personally each member of his community*. It was this change, the shift from the personal to the impersonal society, that was going to cause the human animal its greatest agonies in the millennia ahead. As a species we were not biologically equipped to cope with a mass of strangers masquerading as members of our tribe. It was something we had to learn to do, but it was not easy. As we shall see later, we are still fighting against it today in all kinds of hidden ways—and some that are not so hidden.

As a result of the artificiality of the inflation of human social life to the super-tribal level, it became necessary to introduce more elaborate forms of controls to hold the bulging communities together. The enormous material benefits of super-tribal life had to be paid for in discipline. In the ancient civilizations which began to develop around the

Mediterranean, in Egypt, Greece and Rome and elsewhere, administration and law grew heavier and more complex alongside the increasingly flourishing technologies and arts.

It was a slow process. The magnificence of the remnants of these civilizations that we marvel at today tends to make us think of them as comprising vast populations, but this was not so. In heads per super-tribe, the growth was gradual. As late as 600 B.C., the largest city, Babylon, contained no more than 80,000 people. Classical Athens had a citizen population of only 20,000, and only a quarter of these were members of the true urban elite. The total population of the whole city state, including foreign merchants, slaves and rural as well as urban residents, has been estimated at no more than 70,000 to 100,000. To put this into perspective, it is slightly smaller than present-day university towns such as Oxford or Cambridge. The great modern metropolises, of course, bear no comparison: there are over a hundred today boasting populations exceeding one million, with the biggest exceeding ten million. Modern Athens itself contains no fewer than 1,850,000 people.

If they were to continue to grow in splendour, the ancient urban states could no longer rely on local produce. They had to augment their supplies in one of two ways: by trade or by conquest. Rome did both, but put the emphasis on conquest and carried it out with such devastating administrative and military efficiency that it was able to create the greatest city the world had seen, containing a population approaching half a million, and setting a pattern that was to echo widely down the centuries that followed. These echoes exist today, not only in the brain-straining toil of the organizers, manipulators and creative talents, but also in the increasingly idle, sensation-seeking urban élite, who have grown so numerous that they can easily turn sour with shattering effect and must be kept amused at all costs. In the sophisticated city-dweller of Imperial Rome, we can already see a prototype of the present-day super-tribesman.

In unfolding our urban tale we have, with ancient Rome, come to a stage where the human community has grown so big and is so densely packed that, zoologically speaking, we have already arrived at the modern condition. It is true that, during the centuries that followed, the plot thickened, but it was essentially the same plot. The crowds became denser, the élite became éliter, the technologies became more technical. The frustrations and stresses of city life became greater. Super-tribal clashes became bloodier. There were too many people and that meant there were people to spare, people to waste. As human relationships, lost in the crowd, became ever more impersonal, so man's inhumanity to man increased to horrible proportions. However, as I have said before, an impersonal relationship is not a biologically human one, so this is not surprising. What is surprising is that the bloated super-tribes have survived at all and, what is more, survived so well. This is not something we should accept simply because we are sitting here in the twentieth century, it is something we should marvel at. It is an astonishing testimony to our incredible ingenuity, tenacity and plasticity as a species. How on earth did we manage it? All we had to go on, as animals, was a set of biological characteristics evolved during our long hunting apprenticeship. The answer must lie in the nature of these characteristics and the way we have been able to exploit and manipulate them without distorting them as badly as we (superficially) seem to have done. We must take a closer look at them.

Bearing in mind our monkey ancestry, the social organization of surviving monkey species can provide us with some revealing clues. The existence of powerful, dominant individuals, lording it over the rest of the group, is a widespread phenomenon among higher primates. The weaker members of the group accept their subordinate roles. They do not rush off into the undergrowth and set up on their own. There is strength and security in numbers. When these numbers become too great, then, of course, a splinter group will form and depart, but isolated individual monkeys are abnormalities. The groups move about together and keep together at all times. This allegiance is not merely the result of an enforced tyranny on the part of the leaders, the dominant males. Despots they may be, but they also play another role, that of guardians and protectors. If there is a threat to the group from without, such as an attack from a hungry predator, it is they who are most active in defence. In the face of an external challenge, the top males must get together to meet it, their internal squabbles forgotten. But on other occasions active co-operation within the group is at a minimum.

Returning to the human animals, we can see that this basic system—social co-operation when facing outward, social competition when facing inward—also applies to us, although our early human ancestors were forced to shift the balance somewhat. Their gargantuan struggle to convert from fruit-eaters to hunters required much greater, more active, internal co-operation. The external world, in addition to providing occasional panics, now threw an almost non-stop challenge in the face of the emerging hunter. The result was a basic shift towards mutual aid, towards sharing and combining resources. This does not mean that early man began to move as one, like a shoal of fish; life was too complex for that. Competition and leadership remained, helping to provide impetus and reduce indecisiveness, but despotic authority was severely curtailed. A delicate balance was achieved and, as we have already seen, one that was to prove immensely successful, enabling the early human hunters to spread over most of the earth's surface, with only the minimum of technology to help them on their way.

What happened to this delicate balance as the tiny tribes blossomed into giant super-tribes? With the loss of the person-to-person tribal pattern, the competitive/co-operative pendulum began to swing dangerously back and forth, and it has been oscillating damagingly ever since. Because the subordinate members of the super-tribes became impersonal crowds, the most violent swings of the pendulum have been towards the domineering, competitive side. The overgrown urban groups rapidly and repeatedly fell prey to exaggerated forms of tyranny, despotism and dictatorship. The super-tribes gave rise to super-leaders, exercising powers that make the old monkey tyrants look positively benign. They also gave rise to super-subordinates in the form of slaves, who suffered subservience of a kind more extreme than anything even the most lowly of monkeys would have known.

It took more than a single despot to dominate a super-tribe in this way. Even with deadly new technologies—weapons, dungeons, tortures—to aid him in forcibly maintaining conditions of widespread subjugation, he also required a massive following if he was to succeed in holding the biological pendulum so far to one side. This was possible because the followers, like the leaders, were infected by the impersonality of the super-tribal condition. They calmed their co-operative consciences to some extent by the device of setting up sub-groups, or pseudo-tribes, within the main body of the super-tribe. Each individual established personal relationships of the old, biological type with a small, tribe-sized group of social or professional companions. Within that group he could satisfy his basic urges towards mutual aid and sharing. Other sub-groups—the slave class, for example—could then be looked upon more comfortably as outsiders beyond his protection. The social 'double standard' was born. The insidious strength of these new sub-divisions lay in the fact that they even made it possible for personal relationships to be carried on in an impersonal way. Although a subordinate—a slave, a servant or a serf—might be personally known to a master, the fact that he had been neatly placed into another social category meant that he could be treated as badly as one of the impersonal mob.

It is only a partial truth to say that power corrupts. Extreme subjugation can corrupt equally effectively. When the bio-social pendulum swings away from active co-operation towards tyranny, the whole society becomes corrupt. It may make great material strides. It may shift 4,883,000 tons of stone to build a pyramid; but with its deformed social structure its days are numbered. You can dominate just so much, just so long and just so many, but even within the hothouse atmosphere of a super-tribe, there is a limit. If, when that limit is reached, the bio-social pendulum tilts gently back to its balanced mid-point, the society can count itself lucky. If, as is more likely, it swings wildly back and forth, the blood will flow on a scale our primitive hunting ancestor would never have dreamt of.

It is the miracle of civilized survival that the human co-operative urge reasserts itself so strongly and so repeatedly. There is so much working against it and yet it keeps on coming back. We like to think of this as the conquest of bestial weaknesses by the powers of intellectual altruism, as if ethics and morality were some kind of modern invention. If this were really true, it is doubtful if we would be here today to proclaim it. If we did not carry in us the basic biological urge to co-operate with our fellow men, we would never have survived as a species. If our hunting ancestors had really been ruthless, greedy tyrants, loaded with 'original sin', the human success story would have petered out long ago. The only reason why we are always having the doctrine of original sin instilled into us, in one form or another, is that the artificial conditions of the super-tribe keep on working against our biological altruism, and it needs all the help it can get.

I am aware that there are some authorities who will disagree violently with what I have just said. They see men as naturally inclined to be weak, greedy and wicked, requiring stern codes of imposed control to make them strong, kind and good. But when they deride the concept of the 'noble savage' they confuse the issue. They point out that there was nothing noble about ignorance or superstition, and in that respect they are right. But it is only part of the story. The other part concerns the early hunter's conduct towards his companions. Here the situation must have been different. Compassion, kindness, mutual assistance, a fundamental urge to co-operate within the tribe *must* have been the pattern for the early groups of men to survive in their precarious environment. It was only when the tribes expanded into impersonal super-tribes that the ancient pattern of conduct came under pressure and began to break down. Only then did artificial laws and codes of discipline have to be imposed to correct the balance. If these had been imposed to a degree to match the new pressures, all would have been well; but in the early civilizations men were novices at achieving this delicate balance. They failed repeatedly, with lethal results. We are more expert now, but the system has never been perfected because, as the super-tribes have continued to swell, the problem has re-set itself.

Let me put it in another way. It has often been said that 'the law forbids men to do only what their instincts incline them to do'. It follows from this that if there are laws against theft, murder and rape, then the human animal must, by nature, be a thieving, murderous rapist. Is this really a fair description of man as a social biological species? Somehow, it does not fit the zoological picture of the emerging tribal species. Sadly, however, it does fit the super-tribal picture.

Theft, perhaps the most common of crimes, is a good example. A member of a super-tribe is under pressure, suffering from all the stresses and strains of his artificial social condition. Most people in his super-tribe are strangers to him; he has no personal, tribal bond with them. The typical thief is not stealing from one of his known companions. He is not breaking the old, biological tribal code. In his mind, he is simply setting his victim outside his tribe altogether. To counteract this, a super-tribal law has to be imposed. It is relevant here that we sometimes talk of 'honour among thieves' and the 'code of the underworld'. This underlines the fact that we look upon criminals as belonging to a separate and distinct pseudo-tribe within the super-tribe. In passing, it is interesting to note how we deal with the criminal: we shut him away in a confined, all-criminal community. As a short-term solution it works well enough, but the long-term effect is that it strengthens his pseudo-tribal identity instead of weakening it and furthermore helps him to widen his pseudo-tribal social contacts.

Reconsidering the idea that 'the law forbids men to do only what their instincts incline them to do', we might reword it to the effect that 'the law forbids men to do only what the artificial conditions of civilization drive them to do'. In this way we can see the law as a balancing device, tending to counteract the distortions of super-tribal existence and helping to maintain, in unnatural conditions, the forms of social conduct natural to the human species.

However, this is an over-simplification. It implies perfection in the leaders, the law-makers. Tyrants and despots can, of course, impose harsh and unreasonable laws restraining the population to a greater extent than is justified by the prevailing super-tribal conditions. A weak leadership may impose a system of law that lacks the strength to hold together a teeming populace. Either way lies cultural disaster or decline.

There is also another kind of law that has little to do with the argument I have been putting forward, except that it serves to hold society together. It is an 'isolating law', one which helps to make one culture different from another. It gives cohesion to a society by providing it with a unique identity. These laws play only a minor role in the law courts. They are more the concern of religion and social custom. Their function is to increase the illusion that one belongs to a unified tribe rather than a sprawling, seething super-tribe. If they are criticized because they seem arbitrary or meaningless, the answer comes back that they are traditional and must be obeyed without question. It is as well not to question them because, in themselves, they *are* arbitrary and frequently meaningless. Their value lies in the fact that they are shared by all the members of the community. When they fade, the unity of the community fades a little, too. They take many forms: the elaborate procedures of social ceremonies—marriages, burials, celebrations, parades, festivals and the rest; the intricacies of social etiquette, manners and protocol; the complexities of social costume, uniform decoration, adornment and display.

These subjects have been studied in detail by ethnologists and cultural anthropologists, who have been fascinated by their great diversity. Diversity, the differentiation of one culture from another, has, of course, been the very function of these patterns of behaviour. But in marvelling at their variety, one must not overlook their fundamental similarities. The customs and costumes may be strikingly different *in detail* from culture to culture, but they have the same basic function and the same basic forms. If you start by making a list of all the social customs of one particular culture, you will find equivalents to nearly all of them in nearly all other cultures. Only the details will differ, and they will differ so wildly that they will sometimes obscure the fact that you are dealing with the same basic social patterns.

To give an example: in some cultures ceremonies of mourning involve the wearing of black costumes; in others, by complete contrast, the mourning dress is white. Furthermore, if you cast the net wide enough, you can find still other cultures that employ dark blue, or grey, or yellow, or natural brown sackcloth. Having grown up yourself in a culture where, from early childhood, one of these colours, say black, has been heavily associated with death and mourning, it will be startling to think of wearing such colours as yellow or blue in this context. Therefore, your immediate reaction on discovering that these colours *are* worn as mourning dress in other places is to remark on how different they are from your own familiar custom. Herein lies the trap, so neatly set by the demands of cultural isolation. The superficial observation that the colours vary so dramatically obscures the more basic fact that all these cultures share the performance of a mourning 'display', and that in all of them it involves the wearing of a costume that is strikingly different from the non-mourning costume.

In the same way, when an Englishman first visits Spain, he is surprised to find the public spaces of the towns and villages thronged with people in the early evening, all wandering up and down in an apparently aimless way. His immediate reaction is not that this is their cultural equivalent of his more familiar cocktail parties, but rather that it is some sort of strange local custom. Again, the basic social pattern is the same, but the details differ.

Similar examples could be given to cover almost all forms of communal activity, the principle being that the more social the occasion, the more variable the details and the stranger the other culture's behaviour appears to be at first sight. The greatest social occasions of all, such as coronations, state funerals, balls, banquets, independence celebrations, investitures, great sporting events, military parades, festivals and garden parties (or their equivalents), are the ones where the isolating law plays the strongest part. They vary from case to case in a thousand tiny details, each of which is scrupulously attended to, as though the very lives of the participants depended upon it. In a sense, of course, their social lives do depend on it, for it is only by their conduct in public places that they can strengthen and support their feelings of social identity, of belonging to a cultural group, and the grander the occasion, the stronger the boost.

This is a fact that successful revolutionaries sometimes overlook or underestimate. In ridding themselves of the old power structure that they have come to detest, they are forced to sweep away with it most of the old ceremonials. Even though these ritual procedures may have nothing directly to do with the overthrown power system, they are too strongly reminiscent of it and must go. A few hurriedly improvised performances may be put in their place, but it is difficult to invent rituals overnight. (It is an interesting sidelight on the Christian movement that its early success depended to some extent on its making a take-over bid for many of the old pagan ceremonies and incorporating them, suitably disguised, into their own festive occasions.) When all the excitement and upheaval of the revolution is over, the eventual unhappiness of many a disgruntled post-revolutionary is due, in a concealed way, to his sense of loss of social occasion and pageantry. Revolutionary leaders would do well to anticipate this problem. It is not the chains of social identity that their followers will want to break, it is the chains of a *particular* social identity. As soon as these are smashed, they will need new ones and will soon become dissatisfied merely with an abstract sense of 'freedom'. Such are the demands of the isolating laws.

Other aspects of social behaviour are also brought into play as cohesive forces. Language is one of these. We tend to think of language exclusively as a communication device, but it is more than that. If it were not, we should all be speaking with the same tongue. Looking back through super-tribal history, it is easy to see how the anti-communication function of language has been almost as important as its communication function. More than any social custom, it has set up enormous inter-group barriers. More than anything else, it has identified an individual as a member of a particular super-tribe, and put obstacles in the way of his defecting to another group.

As the super-tribes have grown and merged, so local languages have been merged, or submerged, and the total number in the world is being reduced. But as this happens, a counter-trend develops: accents and dialects become more socially significant, slang, cant and jargon are invented. As the members of a massive super-tribe attempt to strengthen their tribal identities by setting up sub-groups, so a whole spectrum of 'tongues' develops within the official, major language. Just as English and German act as identity badges and isolating mechanisms between an Englishman and a German, so an upper-class English accent isolates its owner from a lower-class one, and the jargon of chemistry and psychiatry isolates chemists from psychiatrists. (It is a sad fact that the academic world which, in its educational role, should be devoted to communication above all else, exhibits pseudo-tribal isolating languages as extreme as criminal slang. The excuse is that precision of expression demands it. This is true up to a point, but the point is frequently and blatantly exceeded.)

The jargon of slang words can become so specialized that it is almost as if a new language is being born. It is typical of slang expressions that once they spread and become common property, they are replaced with new terms by the originating sub-group. If they are adopted by the whole super-tribe and slide into the official language, then they have lost their original function. (It is doubtful whether you are using the same slang expression for, say, an attractive girl, a policeman or a sexual act, that your parents employed when they were your age. But you still use the same official words.) In extreme cases, one sub-group will adopt an entirely foreign language. The Russian court at one point, for example, spoke in French. In Britain one still sees remnants of this kind of behaviour at the more expensive restaurants, where the menus are usually in French.

Religion has operated in much the same way as language, strengthening bonds within a group and weakening them between groups. It acts on the single, simple premise that there are powerful forces at work above and beyond the ordinary human members of the group and that these forces, these super-leaders or gods, must be pleased, appeased and obeyed without question. The fact that they are never around to be questioned helps them to maintain their position.

At first the powers of the gods were limited and their spheres of influence divided, but as the super-tribes swelled to increasingly unmanageable proportions, greater cohesive forces were needed. A government of minor gods was not strong enough. A massive super-tribe required a single, all-powerful, all-wise, all-seeing god, and from the ancient candidates it was this type that won through and survived the passage of the centuries. In the smaller and more backward cultures today the minor gods still rule, but members of all the major cultures have turned to the single super-god.

It is a common observation that the power of religion as a social force has been weakening during recent years. There are two reasons for this. In the first place, it is failing to serve its double function as a cohesive influence. As the populations continued to grow and swell, the ancient empires became unmanageable and split up into national groups. The new super-tribes fought to establish their identities, using all the usual devices. But many of them now shared a common religion. This meant that, for them, religion, although still a powerful force for bonding together the members of a nation, failed in its other cohesive function, namely to weaken the bonds between nations. A compromise was reached by the formation of sects within a major religion. Although sectarianism put back some of the isolating qualities and helped to tribalize, or localize, religious ceremonies again, it was only a partial solution.

The second reason for its loss of power was the growing standard of widespread scientific education, with its increasing demands that the individual *should* ask questions, rather than blindly accept dogmas. The Christian religion, in particular, has suffered serious setbacks. The increasingly logical mind of the Western super-tribesman cannot help but notice certain glaring illogicalities. Perhaps the most important of these is the great discrepancy between, on the one hand, the teaching of humility and gentleness and, on the other, the elaborate finery, pomp and power of the Church leaders.

In addition to law, custom, language and religion there is another, more violent form of cohesive force that helps to bind the members of a super-tribe together, and that is war. To put it cynically, one could say that nothing helps a leader like a good war. It gives him his only chance of being a tyrant and being loved for it at the same time. He can introduce the most ruthless forms of control and send thousands of his followers to their deaths and still be hailed as a great protector. Nothing ties tighter the in-group bonds than an out-group threat.

The fact that internal squabbles are suppressed by the existence of a common enemy has not escaped the attention of rulers past and present. If an overgrown super-tribe is beginning to split at the seams, the splits can rapidly be stitched up by the appearance of a powerful hostile THEM that converts us into a unified US. Just how often leaders have deliberately manipulated an inter-group clash with this in mind, it is hard to say, but whether the move is consciously deliberate or not, the cohesive reaction nearly always occurs. It takes a remarkably inefficient leader to bungle it. Of course, he has to have an enemy who is capable of being painted in sufficiently villainous colours, or he is likely to be in trouble. The disgusting horrors of war become converted into glorious battles only when the threat from outside is really serious, or can be made to seem so.

Despite its attractions to a ruthless leader, war has an obvious disadvantage: one side is liable to be utterly defeated and it might be his. The super-tribesman can be grateful for this unfortunate drawback.

These, then, are the cohesive forces that are brought to bear on the great urban societies. Each has developed its own specialized kind of leader: the administrator, the judge, the politician, the social leader, the high priest, the general. In simpler times they were all rolled into one, an omnipotent emperor or king who was able to cope with the whole range of leadership. But as time has passed and the groups have expanded, the true leadership has shifted from one sphere to another, moving to whichever category happens to contain the most exceptional individual.

In more recent times it has often become the practice to allow the populace to have a say in the election of a new leader. This political device has, in itself, been a valuable cohesive force, giving the super-tribesmen a greater feeling of 'belonging' to their group and having some influence over it. Once the new leader has been elected it soon becomes clear that the influence is slighter than imagined, but nevertheless at the time of the election itself a valuable ripple of social identity runs through the community.

As an aid to this process, local pseudo-tribal sub-leaders are sent to participate in the government of the land. In some countries this has become little more than a ritual act, since the 'local' representatives are no more than imported professionals. However, this type of distortion is inevitable in a community as complex as a modern super-tribe.

The aim of government by elected representatives is fine and clear, even if it is difficult to match in practice. It is based on a partial return to the 'politics' of the original human tribal system, where each member of the tribe (or at least, the adult males) had a say in the running of the society. They were, in a sense, communists, with the emphasis on sharing and with little regard for the rigid protection of personal property. Property was as much for giving as for keeping. But as I have said before, the tribes were small and everyone knew everyone else. They may have prized individual possessions, but doors and locks were things of the future. As soon as the tribe had become an impersonal super-tribe, with strangers in its midst, the rigorous protection of property became necessary and began to play a much larger role in social life. Any political attempts to ignore this fact would meet with considerable difficulties. Modern communism is beginning to find this out and has already started to adjust its system accordingly.

Another adjustment was also necessary in all cases where the aim was to reinstate the ancient tribal-hunter pattern of 'government of the people, by the people'. The super-tribes were simply too big and the problems of government too complex, too technical. The situation demanded a system of representation and it demanded a professional class of experts. Just how far this can get away from 'government by the people' was clearly illustrated in England recently when it was suggested that parliamentary debates should be televised so that, thanks to modern science, the populace could at last play a more intimate role in the affairs of state. But this would have interfered with the specialized, professional atmosphere and was hotly opposed and rejected. So much for government by the people. However, this is not surprising. Running a super-tribe is like trying to balance an elephant on a tight-rope. It seems that the best that any modern political system can hope for is to use right-wing methods to implement left-wing policies. (This is, in effect, what is being done both in the East and the West at the present moment.) It is a difficult trick to pull off and requires great professional finesse, not to mention double-talk. If modern politicians are frequently the subject of scorn and satire, it is because too many people see through the trick too often. But given the size of the present super-tribes there appears to be no alternative.

Because the modern super-tribes are in so many ways socially unmanageable, there has been a great tendency for them to fragment. I have already mentioned the way in which specialized pseudo-tribes crystallize inside the main body, as social groups, class groups, professional groups, academic groups, sports groups and so on, reinstating for the urban individual various forms of tribal identity. These groups remain happily enough within the main community, but more drastic splits than this frequently occur. Empires split into independent countries; countries split into self-ruling sectors. Despite improved communications, despite increasingly shared aims and common policies, the splits go on. Alliances can be quickly forged under the cohesive pressure of war, but during peace, separations and divisions are the order of the day. When splinter-groups desperately struggle to forge some kind of local identity, it simply means that the cohesive forces of the super-tribe to which they belonged were not strong enough or exciting enough to hold them together.

The dream of a peaceful, global super-tribe is repeatedly being shattered. It seems as if only an alien threat from another planet would provide the necessary cohesive force, and then only temporarily. It remains to be seen whether, in the future, man's ingenuity will introduce some new factor into his social existence that will solve the problem. At the moment it appears unlikely.

There has been a great deal of debate recently concerning the way in which modern mass-communication devices, such as television, are 'shrinking' the social surface of the world, creating a global televillage. It has been suggested that this trend will aid the move towards a genuinely international community. Unhappily this is a myth, for the single reason that television, unlike personal social intercourse, is a *one-way* system. I can listen to and get to know a tele-speaker, but he cannot listen to or get to know me. True, I can learn what he is thinking and doing and this is admittedly a great advantage, broadening my range of social information, but it is no substitute for the two-way relationships of real social contacts.

Even if startlingly new and at present unimagined advances in mass-communication techniques are made in the years to come, they will continue to be hampered by the bio-social limitations of our species. We are not equipped, like termites, to become willing members of a vast community. We are and probably always will be, at base, simple tribal animals.

Yet despite this, and despite the spasmodic fragmentations that are constantly occurring around the globe, we are bound to face the fact that the major trend is still to maintain the massive super-tribal levels. While splits are occurring in one part of the world, mergers are developing in another. If the situation remains as unstable today as it has been for centuries, then why do we persist with it? If it is so dangerous, why do we keep it up?

It is far more than just an international power game. There is an intrinsic, biological property of the human animal that obtains deep satisfaction from being thrown into the urban chaos of a super-tribe. That quality is man's insatiable curiosity, his inventiveness, his intellectual athleticism. The urban turmoil seems to energize this quality. Just as colony-nesting sea-birds are reproductively aroused by massing in dense breeding communities, so the human animal is intellectually aroused by massing in dense urban communities. They are breeding colonies of human ideas. This is the credit side of the story. It keeps the system going despite its many disadvantages.

We have looked at some of these disadvantages on the social level, but they exist on the personal level as well. Individuals living in a large urban complex suffer from a variety of stresses and strains: noise, polluted air, lack of exercise, cramping of space, overcrowding, over-stimulation and, paradoxically, for some, isolation and boredom.

You may think that the price the super-tribesman is paying is too high; that a quiet, peaceful, contemplative life would be far preferable. He thinks so, too, of course, but like that physical exercise he is always going to take, he seldom does anything about it. Moving to the suburbs is about as far as he goes. There he can create a pseudo-tribal atmosphere away from the strains of the big city, but come Monday morning and he is dashing back into the fray again. He could move away, but he would miss the excitement, the excitement of the neo-hunter, setting off to capture the biggest game in the biggest and best hunting grounds his environment has to offer.

On this basis one might expect every great city to be a raging inferno of novelty and inventiveness. Compared with a village it may seem to be so, but it is very far from reaching its exploratory limits. This is because there is a fundamental clash between the cohesive and inventive forces of society. The one tends to keep things steady and therefore repetitive and static. The other strains on to new developments and the inevitable rejection of old patterns. Just as there is a conflict between competition and co-operation, so there is a fight between conformity and innovation. Only in the city does sustained innovation stand a real chance. Only the city is strong enough and secure enough in its amassed conformity to tolerate the disruptive forces of rebellious originality and creativity. The sharp swords of iconoclasm are mere pin-pricks in the giant's flesh, giving it a pleasant tingling sensation, rousing it from sleep and urging it into action.

This exploratory excitement, then, with the help of the cohesive forces I have described, is what keeps so many modern city-dwellers voluntarily locked inside their human zoo cages. The exhilarations and challenges of super-tribal living are so great that with a little assistance they can outweigh the enormous dangers and disadvantages. But how do the drawbacks measure up to those of the animal zoo?

The animal zoo inmate finds itself in solitary confinement, or in an abnormally distorted social group. Alongside, in other cages, it may be able to see or hear other animals but it cannot make any real contact with them. Ironically, the super-social conditions of human urban life can work in much the same way. The loneliness of the city is a well-known hazard. It is easy to become lost in the great impersonal crowd. It is easy for natural family groupings and personal tribal relationships to become distorted, crushed or fragmented. In a village all the neighbours are personal friends or, at worst, personal enemies; none are strangers. In a large city many people do not even know the names of their neighbours.

This de-personalizing does help to support the rebels and innovators who, in a smaller, tribal community, would be subjected to much greater cohesive forces. They would be flattened by the demands of conformity. But at the same time the paradox of the social isolation of the teeming city can cause a great deal of stress and misery for many of the human zoo inmates.

Apart from the personal isolation there is also the direct pressure of physical crowding. Each kind of animal has evolved to exist in a certain amount of living space. In both the animal zoo and the human zoo this space is severely curtailed and the consequences can be serious. We think of claustrophobia as an abnormal response. In its extreme form it is, but in a milder, less clearly recognized form it is a condition from which all city-dwellers suffer. Half-hearted attempts are made to correct this. Special sections of the city are set aside as a token gesture towards providing open spaces—small bits of 'natural environment' called parks. Originally parks were hunting grounds containing deer and other prey species, where rich super-tribesmen could re-live their ancestral patterns of hunting behaviour; but in modern city parks only the plant life remains.

In terms of quantity of space, the city park is a joke. It

would have to cover thousands of square miles to provide a truly natural amount of wandering space for the huge city population it serves. The best that can be said for it is that it is decidedly better than nothing.

The alternative for the urban space-seekers is to make brief sorties into the countryside, and this they do with great vigour. Bumper to bumper the cars set off each week-end, and bumper to bumper they return. But no matter, they have wandered—they have patrolled a broader home range—and in so doing have kept up the fight against the unnatural spatial cramping of the city. If the crowded roads of the modern super-tribe have turned this into something of a ritual, it is still preferable to giving up. The position is even worse for the inmates of the animal zoo. Their version of bumper-to-bumper patrolling is the even more stultified pacing to and fro across their cage floor. But they do not give up either. We should be thankful that we can do more than pace back and forth across our living-room floors.

Having now traced the course of events that has led us to our present social condition, we can start to examine in more detail the various ways in which our behaviour patterns have succeeded in adjusting to life in the human zoo, or, in some instances, how they have disastrously failed to do so.

CHAPTER TWO

STATUS AND SUPER-STATUS

In any organized group of mammals, no matter how co-operative, there is always a struggle for social dominance. As he pursues this struggle, each adult individual acquires a particular social rank, giving him his position, or status, in the group hierarchy. The situation never remains stable for very long, largely because all the status strugglers are growing older. When the overlords, or 'top-dogs', become senile, their seniority is challenged and they are overthrown by their immediate subordinates. There is then renewed dominance squabbling as everyone moves a little farther up the social ladder. At the other end of the scale, the younger members of the group are maturing rapidly, keeping up the pressure from below. In addition, certain members of the group may suddenly be struck down by disease or accidental death, leaving gaps in the hierarchy that have to be quickly filled.

The general result is a constant condition of *status tension*. Under natural conditions this tension remains tolerable because of the limited size of the social groupings. If, however, in the artificial environment of captivity, the group size becomes too big, or the space available too small, then the status 'rat race' soon gets out of hand, dominance battles rage uncontrollably, and the leaders of the packs, prides, colonies or tribes come under severe strain. When this happens, the weakest members of the groups are frequently hounded to their deaths, as the restrained rituals of display and counter-display degenerate into bloody violence.

There are further repercussions. So much time has to be spent sorting out the unnaturally complex status relationships that other aspects of social life, such as parental care, become seriously and damagingly neglected.

If the settling of dominance disputes creates difficulties for the moderately crowded inmates of the animal zoo, then it is obviously going to provide an even greater dilemma for the vastly overgrown super-tribes of the human zoo. The essential feature of the status struggle in nature is that it is based on the *personal* relationships of the individuals inside the social group. For the primitive human tribesman the problem was therefore a comparatively simple one, but when the tribes grew into super-tribes and relationships became increasingly impersonal, the problem of status rapidly expanded into the nightmare of super-status.

Before we probe this tender area of urban life, it will be helpful to take a brief look at the basic laws which govern the dominance struggle. The best way to do this is to survey the battlefield from the viewpoint of the dominant animal.

If you are to rule your group and to be successful in holding your position of power, there are ten golden rules you must obey. They apply to all leaders, from baboons to modern presidents and prime ministers. The ten commandments of dominance are these:

1. *You must clearly display the trappings, postures and gestures of dominance.*

For the baboon this means a sleek, beautifully groomed, luxuriant coat of hair; a calm, relaxed posture when not engaged in disputes; a deliberate and purposeful gait when active. There must be no outward signs of anxiety, indecision or hesitancy.

With a few superficial modifications, the same holds true for the human leader. The luxuriant coat of fur becomes the rich and elaborate costume of the ruler, dramatically excelling those of his subordinates. He assumes postures unique to his dominant role. When he is relaxing, he may recline or sit, while others must stand until given permission to follow suit. This is also typical of the dominant baboon, who may sprawl out lazily while his anxious subordinates hold themselves in more alert postures near by. The situation changes once the leader stirs into aggressive action and begins to assert himself. Then, be he baboon or prince, he must rise into a more impressive position than that of his followers. He must literally rise above them, matching his psychological status with his physical posture. For the baboon boss this is easy: a dominant monkey is nearly always much larger than his underlings. He has only to hold himself erect and his greater body size does the rest. The situation is enhanced by cringing and crouching on the part of his more fearful subordinates. For the human leader, artificial aids may be necessary. He can magnify his size by wearing large cloaks or tall headgear. His height can be increased by mounting a throne, a platform, an animal, or a vehicle of some kind, or by being carried aloft by his followers. The crouching of the weaker baboons becomes stylized in various ways: subordinate humans lower their height by bowing, curtsying, kneeling, kowtowing, salaaming or prostrating.

The ingenuity of our species permits the human leader to have it both ways. By sitting on a throne on a raised platform, he can enjoy both the relaxed position of the passive dominant *and* the heightened position of the active dominant at one and the same time, thus providing himself with a doubly powerful display posture.

The dignified displays of leadership that the human animal shares with the baboon are still with us in many forms today. They can be seen in their most primitive and obvious conditions in generals, judges, high priests and surviving royalty. They tend to be more limited to special occasions than they once were, but when they do occur they are as ostentatious as ever. Not even the most learned academics are immune to the demands of pomp and finery on their more ceremonial occasions.

Where emperors have given way to elected presidents and prime ministers, personal dominance displays have, however, become less overt. There has been a shift of emphasis in the role of leadership. The new-style leader is a servant of the people who happens to be dominant, rather than a dominator of the people who also serves them. He underlines his acceptance of this situation by wearing a comparatively drab costume, but this is only a trick. It is a minor dishonesty that he can afford, to make him seem more 'one of the crowd', but he dare not carry it too far or, before he knows it, he really will have become one of the crowd again. So, in other, less blatantly personal ways, he must continue to perform the outward display of his dominance. With all the complexities of the modern urban environment at his disposal, this is not difficult. The loss of grandeur in his dress can be compensated for by the elaborate and exclusive nature of the rooms in which he rules and the buildings in which he lives and works. He can retain ostentation in the way he travels, with motorcades, out-riders and personal planes. He can continue to surround himself with a large group of 'professional subordinates'—aides, secretaries, servants, personal assistants, bodyguards, attendants and the rest—part of whose job is merely to be seen to be servile towards him, thereby adding to his image of social superiority. His postures, movements and gestures of dominance can be retained unmodified. Because the power signals they transmit are so basic to the human species, they are accepted unconsciously and can therefore escape restriction. His movements and gestures are calm and relaxed, or firm and deliberate. (When did you last see a president or a prime minister running, except when taking voluntary exercise?) In conversation he uses his eyes like weapons, delivering a fixed stare at moments when subordinates would be politely averting their gaze, and turning his head away at moments when subordinates would be watching intently. He does not scrabble, twitch, fidget or falter. These are essentially the reactions of subordinates. If the leader performs them there is something seriously wrong with him in his role as the dominant member of the group.

2. *In moments of active rivalry you must threaten your subordinates aggressively.*

At the slightest sign of any challenge from a subordinate baboon, the group leader immediately responds with an impressive display of threatening behaviour. There is a whole range of threat displays available, varying from those motivated by a lot of aggression tinged with a little fear to those motivated by a lot of fear and only a little aggression. The latter—the 'scared threats' of weak-but-hostile individuals—are never shown by a dominant animal unless his leadership is tottering. When his position is secure he shows only the most aggressive threat displays. He can be so secure that all he needs to do is to indicate that he is about to threaten, without actually bothering to carry it through. A mere jerk of his massive head in the direction of the unruly subordinate may be sufficient to subdue the inferior individual. These actions are called 'intention movements', and they operate in precisely the same way in the human species. A powerful human leader, irritated by the actions of a subordinate, need only jerk his head in the latter's direction and fix him with a hard stare, to assert his dominance successfully. If he has to raise his voice or repeat an order, his dominance is slightly less secure, and he will, on eventually regaining control, have to re-establish his status by administering a rebuke or a symbolic punishment of some kind.

The act of raising his voice, or raging, is only a weak sign in a leader when it occurs as a reaction to an immediate threat. It may also be used spontaneously or deliberately by a strong ruler as a general device for reaffirming his position. A dominant baboon may behave in the same way, suddenly charging at his subordinates and terrorizing them, reminding them of his powers. It enables him to chalk up a few points, and after that he can more easily get his own way with the merest nod of his head. Human leaders perform in this manner from time to time, issuing stern edicts, making lightning inspections or haranguing the group with vigorous speeches. If you are a leader, it is dangerous to remain silent, unseen or unfelt for too long. If natural circumstances do not prompt a show of power, the circumstances must be invented that do. It is not enough to have power, one must be observed to have power. Therein lies the value of spontaneous threat displays.

3. *In moments of physical challenge you (or your delegates) must be able forcibly to overpower your subordinates.*

If a threat display fails, then a physical attack must follow. If you are a baboon boss this is a dangerous step to take, for two reasons. Firstly, in a physical fight even the winner may be damaged, and injury is more serious for a dominant animal than for a subordinate. It makes him less daunting for a subsequent attacker. Secondly, he is always outnumbered by his subordinates, and if they are driven too far they may gang up on him and overpower him in a combined effort. It is these two facts that make threat rather than actual attack the preferred method for dominant individuals.

The human leader overcomes this to some extent by employing a special class of 'suppressors'. They, the military or police, are so specialized and professional at their task that only a general uprising of the whole populace would be strong enough to beat them. In extreme cases, a despot will employ a further, even more specialized class of suppressors (such as secret police), whose job it is to suppress the ordinary suppressors if they happen to get out of line. By clever manipulation and administration it is possible to run an aggressive system of this kind in such a way that only the leader knows enough of what is happening to be able to control it. Everyone else is in a state of confusion unless they have orders from above and, in this way, the modern despot can hold the reins and dominate effectively.

4. *If a challenge involves brain rather than brawn you must be able to outwit your subordinates.*

The baboon boss must be cunning, quick and intelligent as well as strong and aggressive. This is obviously even more important for a human leader. In cases where there is a system of inherited leadership, the stupid individual is quickly deposed or becomes the mere figurehead and pawn of the true leaders.

Today the problems are so complex that the modern leader is forced to surround himself with intellectual specialists, but despite this he cannot escape the need for quick-wittedness. It is he who must make the final decisions, and make them sharply and clearly, without faltering. This is such a vital quality in leadership that it is more important to make a firm, unhesitating decision than it is to make the 'right' one. Many a powerful leader has survived occasional wrong decisions, made with style and forcefulness, but few have survived hesitant indecisiveness. The golden rule of leadership here, which in a rational age is an unpleasant one to accept, is that it is the manner in which you do something that really counts, rather than what you do. It is a sad truth that a leader who does the wrong things in the right way will, up to a certain point, gain greater allegiance and enjoy more success than one who does the right things in the wrong way. The progress of civilization has repeatedly suffered as a result of this. Lucky indeed is the society whose leader does the right things and at the same time obeys the ten golden rules of dominance; lucky—and rare, too. There appears to be a sinister, more-than-chance relationship between great leadership and aberrant policies.

It seems as if one of the curses of the immense complexity of the super-tribal condition is that it is almost impossible to make sharp, clear-cut decisions, concerning major issues, on a rational basis. The evidence available is so complicated, so diverse and frequently so contradictory, that any reasonable, rational decision is bound to involve undue hesitancy. The great super-tribal leader cannot enjoy the luxury of ponderous restraint and 'further examination of the facts' so typical of the great academic. The biological nature of his role as a dominant animal forces him to make a snap decision or lose face.

The danger is obvious: the situation inevitably favours,

as great leaders, rather abnormal individuals, fired by some kind of obsessive fanaticism, who will be prepared to cut through the mass of conflicting evidence that the super-tribal condition throws up. This is one of the prices that the biological tribesman must pay for becoming an artificial super-tribesman. The only solution is to find a brilliant, rational, balanced, deep-thinking brain housed in a glamorous, flamboyant, self-assertive, colourful personality. Contradictory? Yes. Impossible? Perhaps; but there is a glimmer of hope in the fact that the very size of the super-tribe, which causes the problem in the first place, also offers literally millions of potential candidates.

5. *You must suppress squabbles that break out between your subordinates.*

If a baboon leader sees an unruly squabble taking placc he is likely to interfere and suppress it, even though it does not in any way constitute a direct threat to himself. It gives him another opportunity of displaying his dominance and at the same time helps to maintain order inside the group. Interference of this kind from the dominant animal is directed particularly at squabbling juveniles, and helps to instil in them, at an early age, the idea of a powerful leader in their midst.

The equivalent of this behaviour for the human leader is the control and administration of the laws of his group. The rulers of the earlier and smaller super-tribes were powerfully active in this respect, but there has been increasing delegation of these duties in modern times, due to the increasing weight of other burdens that relate more directly to the status of the leader. Nevertheless, a squabbling community is an inefficient one and some degree of control and influence has to be retained.

6. *You must reward your immediate subordinates by permitting them to enjoy the benefits of their high ranks.*

The sub-dominant baboons, although they are the leader's worst rivals, are also of great help to him in times of threat from outside the group. Further, if they are too strongly suppressed they may gang up on him and depose him. They therefore enjoy privileges which the weaker members of the group cannot share. They have more freedom of action and are permitted to stay closer to the dominant animal than are the junior males.

Any human leader who has failed to obey this rule has soon found himself in difficulties. He needs more help from his sub-dominants, and is in greater danger of a 'palace revolt', than his baboon equivalent. So much more can go on behind his back. The system of rewarding the sub-dominants requires brilliant expertise. The wrong sort of reward gives too much power to a serious rival. The trouble is that a true leader cannot enjoy true friendship. True friendship can only be fully expressed between members of roughly the same status level. A partial friendship can, of course, occur between a dominant and a subordinate, at any level, but it is always marred by the difference in rank. No matter how well meaning the partners in such a friendship may be, condescension and flattery inevitably creep in to cloud the relationship. The leader, at the very peak of the social pyramid, is, in the full sense of the word, permanently friendless; and his partial friends are perhaps more partial than he likes to think. As I said, the giving of favours requires an expert hand.

7. *You must protect the weaker members of the group from undue persecution.*

Females with young tend to cluster around the dominant male baboon. He meets any attack on these females or on unprotected infants with a savage onslaught. As a defender of the weak he is ensuring the survival of the future adults of the group. Human leaders have increasingly extended their protection of the weak to include also the old, the sick and the disabled. This is because efficient rulers not only need to defend the growing children, who will one day swell the ranks of their followers, but also need to reduce the anxieties of the active adults, all of whom are threatened with eventual senility, sudden sickness or possible disability. With most people the urge to give aid in such cases is a natural development of their biologically co-operative nature. But for the leaders it is also a question of making people work more efficiently by taking a serious weight off their minds.

8. *You must make decisions concerning the social activities of your group.*

When the baboon leader decides to move, the whole group moves. When he rests, the group rests. When he feeds, the group feeds. Direct control of this kind is, of course, lost to the leader of a human super-tribe, but he can nevertheless play a vital role in encouraging the more abstract directions his group takes. He may foster the sciences or push towards a greater military emphasis. As with the other golden rules of leadership, it is important for him to exercise this one even when it does not appear to be strictly necessary. Even if a society is cruising happily along on a set and satisfactory course, it is vital for him to change that course in certain ways in order to make his impact felt. It is not enough simply to alter it as a reaction to something that is going wrong. He must spontaneously, of his own volition, insist on new lines of development, or he will be considered weak and colourless. If he has no ready-made preferences and enthusiasms, he must invent them. If he is seen to have what appear to be strong convictions on certain matters, he will be taken more seriously on *all* matters. Many modern leaders seem to overlook this and their political 'platforms' are desperately lacking in originality. If they win the battle for leadership it is not because they are more inspiring than their rivals but simply because they are less uninspiring.

9. *You must reassure your extreme subordinates from time to time.*

If a dominant baboon wishes to approach a subordinate peacefully, it may have difficulty doing so, because its close proximity is inevitably threatening. It can overcome this by performing a reassurance display. This consists of a very gentle approach, with no sudden or harsh movements, accompanied by facial expressions (called lip-smacking) which are typical of friendly subordinates. This helps to calm the fears of the weaker animal and the dominant one can then come near.

Human leaders, who may be characteristically tough and unsmiling with their immediate subordinates, frequently adopt an attitude of friendly submissiveness when coming into personal contact with their extreme subordinates. Towards them they offer a front of exaggerated courtesy, smiling, waving, shaking hands interminably and even fondling babies. But the smiles soon fade as they turn away and disappear back inside their ruthless world of power.

10. *You must take the initiative in repelling threats or attacks arising from outside your group.*

It is always the dominant baboon that is in the forefront of the defence against an attack from an external enemy. He plays the major role as the protector of the group. For the baboon, the enemy is usually a dangerous member of another species, but for the human leader it takes the form of a rival group of the same species. At such moments, his leadership is put to a severe test, but, in a sense, it is less severe than during times of peace. The external threat, as I pointed out in the last chapter, has such a powerful cohesive effect on the members of the threatened group that the leader's task is in many ways made easier. The more daring and reckless he is, the more fervently he seems to be protecting the group who, caught up in the emotional fray, never dare question his actions (as they would in peace-time), no matter how irrational these actions may be. Carried along on the grotesque tidal-wave of enthusiasm that war churns up, the strong leader comes into his own. With the greatest of ease he can persuade the members of his group, deeply conditioned as they are to consider the killing of another human being as the most hideous crime known, to commit this same action as an act of honour and heroism. He can hardly put a foot wrong, but if he does, the news of his blunder can always be suppressed as bad for national morale. Should it become public, it can still be put down to bad luck rather than bad judgment. Bearing all this in mind, it is little wonder that, in times of peace, leaders are prone to invent, or at least to magnify, threats from foreign powers that they can then cast in the role of potential enemies. A little added cohesion goes a long way.

These, then, are the patterns of power. I should make it

clear that I am not implying that the dominant baboon/human ruler comparison should be taken as meaning that we evolved from baboons, or that our dominance behaviour evolved from theirs. It is true that we shared a common ancestor with baboons, way back in our evolutionary history, but that is not the point. The point is that baboons, like our early human forbears, have moved out of the lush forest environment into the tougher world of the open country, where tighter group control is necessary. Forest-living monkeys and apes have a much looser social system; their leaders are under less pressure. The dominant baboon has a more significant role to play and I selected him as an example for this reason. The value of the baboon/human comparison lies in the way it reveals the very basic nature of human dominance patterns. The striking parallels that exist enable us to view the human power game with a fresh eye and see it for what it is: a fundamental piece of animal behaviour. But we must leave the baboons to their simpler tasks and take a closer look at the complications of the human situation.

For the modern human leader there are clearly difficulties in performing his dominant role efficiently. The grotesquely inflated power which he wields means that there is the ever present danger that only an individual with an equally grotesquely inflated ego will successfully be able to hold the super-tribal reins. Also, the immense pressures will easily push him into initiating acts of violence, an all-too-natural response to the strains of super-status. Furthermore, the absurd complexity of his task is bound to absorb him to such an extent that it inevitably makes him remote from the ordinary problems of his followers. A good tribal leader knows exactly what is happening in every corner of his group. A super-tribal leader, hopelessly isolated by his lofty position of super-status, and totally preoccupied by the machinery of power, rapidly becomes cut off.

It has been said that to be a successful leader in the modern world a man has to be prepared to make major decisions with the minimum of information. This is a frightening way to run a super-tribe, and yet it happens all the time. There is too much information available for any one individual to assimilate and, in addition, there is a great deal more, hidden in the super-tribal labyrinth, that can never be made available. A rational solution is to do away with the powerful leader-figure, to relegate him to the ancient, tribal past where he belonged, and to replace him with a computer-fed organization of interdependent, specialized experts.

Something approaching such an organization already exists, of course, and in England any civil servant will tell you without hesitation that it is the civil service that really runs the country. To emphasize his point he will inform you that when parliament is in session his work is seriously hampered; only during parliamentary recesses can serious progress be made. All this is very logical, but unfortunately it is not bio-logical, and the country he claims to be running happens to be made up of biological specimens—the super-tribesmen. True, a super-tribe needs super-control, and if it is too much for one man it might seem reasonable to solve the problem by converting a power-figure into a power-organization. This does not, however, satisfy the biological demands of the followers. They may be able to reason super-tribally, but their feelings are still tribal, and they will continue to demand a real leader in the form of an identifiable, solitary individual. It is a fundamental pattern of their species, and there is no avoiding it. Institutions and computers may be valuable servants to the masters, but they can never themselves become masters (science fiction stories notwithstanding). A diffuse organization, a faceless machine, lacks the essential properties: it cannot inspire and it cannot be deposed. The single dominant human is therefore doomed to struggle on, behaving publicly like a tribal leader, with panache and assurance, while in private he grapples laboriously with the almost impossible tasks of super-tribal control.

Despite the great burdens of present-day leadership and despite the daunting fact that an ambitious male member of a modern super-tribe has a chance smaller than one in a million of becoming the dominant individual of his group, there has been no observable lessening of desire to achieve high status. The urge to climb the social ladder is too ancient, too deeply ingrained to be weakened by a rational assessment of the new situation.

Throughout the length and breadth of our massive communities there are, then, hundreds of thousands of frustrated, would-be leaders with no real hope of leading. What happens to their thwarted ladder-climbing? Where does all the energy go? They can, of course, give up and drop out, but this is a depressing condition. The flaw in the social drop-out's solution is that he does not really drop out at all: he stays put and pours scorn on the rat race that surrounds him. This unhappy state is avoided by the great majority of the super-tribesmen by the simple device of competing for leadership in specialized sub-groups of the super-tribe. For some this is easier than others. A competitive profession or craft automatically provides its own social hierarchy. But even there the odds against achieving true leadership may be too great. This gives rise to the almost arbitrary invention of new sub-groups where competition may prove more rewarding. All kinds of extraordinary cults are set up—everything from canary-breeding and train-spotting to UFO-watching and body-building. In each case the overt nature of the activity is comparatively unimportant. What is really important is that the pursuit provides a new social hierarchy where one did not exist before. Inside it a whole range of rules and procedures is rapidly developed, committees are formed and—most important of all—leaders emerge. A champion canary-breeder or body-builder would, in all probability, have no chance whatsoever of enjoying the heady fruits of dominance, were it not for his involvement in his specialized sub-group.

In this way the would-be leader can fight back against the depressingly heavy social blanket that falls over him as he struggles to rise in his massive super-tribe. The vast majority of all sports, pastimes, hobbies and 'good works' have as their principal function not their specifically avowed aims, but the much more basic aim of follow-the-leader-and-beat-him-if-you-can. However, this is a description and not a criticism. In fact, the situation would be much more grave if this multitude of harmless sub-groups, or pseudo-tribes, did not exist. They funnel off a great deal of the frustrated ladder-climbing that might otherwise cause considerable havoc.

I have said that the nature of these activities is of little significance, but nevertheless it is intriguing to notice how many sports and hobbies involve an element of ritualized aggression, over and above simple competitiveness. To take a single example: the act of 'taking aim' is, in origin, a typically aggressive pattern of co-ordination. It reappears, suitably transformed, in a whole range of pastimes, including bowling, billiards, darts, table tennis, croquet, archery, squash, netball, cricket, tennis, football, hockey, polo, shooting and spear-fishing. In children's toys and fairgrounds it abounds. In a slightly heavier disguise it accounts for a great deal of the appeal of amateur photography: we 'shoot' film, 'capture' on celluloid, take snap 'shots', and our cameras = pistols, rolls of film = bullets, cameras with long telescopic lenses = rifles, and cine-cameras = machine-guns. However, although these symbolic equations may be helpful, they are by no means essential in the search for 'pastime dominance'. Matchbox-top collecting will do almost as well, providing, of course, you can make contact with suitable rivals, similarly preoccupied, whose matchbox-top collections you can then seek to dominate.

The setting up of specialist sub-groups is not the only solution to the super-status dilemma. Localized geographical pseudo-tribes also exist. Each village, town, city and county within a super-tribe develops its own regional hierarchy, providing further substitutes for thwarted super-tribal leadership.

On a smaller scale still, each individual has his own closely knit 'social circle' of personal acquaintances. The list of non-commercial names in his private phone-book or address-book gives a good indication of the extent of this kind of pseudo-tribe. It is particularly important because, as in a true tribe, all its members are personally known to him. Unlike a true tribe, however, all the members are not necessarily known to one another. The social groups overlap and interlock with one another in a complex network. Nevertheless, for each individual his social pseudo-tribe provides one more sphere in which he can assert himself and express his leadership.

Another major super-tribal pattern that has helped to split the group up without destroying it has been the system of social classes. These have existed in much the same basic form from the times of the earliest civilizations: an upper or ruling class, a middle class comprising merchants and specialists and a lower class of peasants and labourers. Sub-divisions have appeared as the groups have swollen, and the details have varied, but the principle has remained the same.

The recognition of distinct classes has made it possible for members of classes below the top one to strive for a more realistic dominance status at their particular class level. Belonging to a class is much more than a mere question of money. A man at the top of his social class may earn more than a man at the bottom of the class above. The rewards of being dominant at his own level may be such that he has no wish to abandon his class-tribe. Overlaps of this sort indicate just how strongly tribal the classes can become.

The class-tribe system of splitting up the super-tribe has, however, suffered serious set-backs in recent years. As the super-tribes grew to even bigger proportions and technologies became more and more complex, so the standard of mass education had to be raised to keep pace with the situation. Education, combined with improvements in mass communication and especially the pressures of mass advertising, led to a major breakdown in class barriers. The comforts of 'knowing your own station' in life were replaced by the exciting and increasingly real possibilities of exceeding that station. Despite this, the old class-tribe system kept on fighting back and is still doing so. We can see the outward signs of this running battle very clearly today in the ever-increasing speed of fashion cycles. New styles of clothing, furniture, decoration, music and art replace one another more and more quickly. It is often suggested that this is the result of commercial interests and pressures, but it would be just as easy—easier, in fact—to go on selling new variations of old themes rather than introducing new themes. Yet new themes are continually demanded, because the old ones permeate so rapidly through the social system. The quicker they reach the lower strata, the quicker they must be replaced by something new and exclusive at the top. History has never before witnessed such an incredible turnover in styles and tastes. The result, of course, is a major loss of the pseudo-tribal identity provided by the old, rigid social class system.

Replacing this loss, to some extent, is a new super-tribal splitting system that has recently developed. Age classes are emerging. A widening gap has appeared between what we must now call a young-adult pseudo-tribe and an old-adult pseudo-tribe. The former possesses its own customs and its own dominance system which are increasingly distinct from those of the latter. The entirely new phenomenon of powerful teenage idols and student leaders has ushered in a major new pseudo-tribal division. Desultory attempts on the part of the old-adult pseudo-tribe to encompass the new group have met with very limited success. The piling of old-adult honours on the heads of young-adult leaders, or the tolerant acceptance of the extremes of young-adult fashions and styles, has only led to further rebellious excesses. (If cannabis smoking is ever legalized and widely adopted, for example, an immediate replacement will be required, just as alcohol had to be replaced by cannabis itself.) When these excesses reach a point that the old adults cannot engulf, or refuse to copy, then the young adults can rest easy for a while. Safely flying their new pseudo-tribal flags, they can enjoy the satisfactions of their new pseudo-tribal independence and their more manageable, self-contained dominance system.

The sobering lesson to be learnt from all this is that the ancient biological need of the human species for a distinct tribal identity is a powerful force that cannot be subdued. As fast as one super-tribal split is invisibly mended, another one appears. Well-meaning authorities talk airily about 'hopes for a global society'. They see clearly the technical possibility of such a development, given the marvels of modern communication, but they stubbornly overlook the biological difficulties.

A pessimistic view? Certainly not. The prospects will remain gloomy only as long as there is a failure to come to terms with the biological demands of the species. Theoretically there is no good reason why small groupings, satisfying the requirements of tribal identity, should not be constructively interrelated inside thriving super-tribes which, in turn, constructively interact to form a massive, global mega-tribe. Failures to date have largely been due to attempts to *suppress* the existing differences between the various groups, rather than to improve the nature of these differences by converting them into more rewarding and peaceful forms of competitive social interaction. Attempts to iron out the whole world into one great expanse of uniform monotony are doomed to disaster. This applies at all levels, from break-away nations to tear-away gangs. When the sense of social identity is threatened, it fights back. The fact that it has to fight for its existence means, at the least, social upheaval and, at worst, bloodshed. We shall be taking a closer look at this in a later chapter, but for the moment we must return to the question of social status and examine it at the level of the individual.

Where exactly does he stand, this modern status-seeker? First, he has his personal friends and acquaintances. Together they form his social pseudo-tribe. Second, he has his local community—his regional pseudo-tribe. Third, he has his specializations: his profession, craft or employment, and his pastimes, hobbies or sports. They make up his specialist pseudo-tribes. Fourth, he has the remnants of a class-tribe and a new age-tribe.

Put together, these sub-groupings provide him with a much greater chance of achieving some sort of dominance and of satisfying his basic status urge, than if he were simply a tiny unit in a homogeneous mass, a human ant crawling about in a gigantic, super-tribal ant-hill. So far, so good; but there are snags.

To begin with, the dominance achieved in a limited sub-group is itself limited. It may be real, but it is only a partial solution. It is impossible to ignore the fact that there are bigger things going on all around. Being a big fish in a little pond cannot blot out dreams of a bigger pond. In the past this was not such a problem, because the rigid class system, ruthlessly applied, kept everyone in his 'place'. This may have been very neat, but it could all too easily lead to super-tribal stagnation. Individuals with minor talents were well served, but many of those with greater talents were held back, frittering their energies away on strictly limited goals. It was possible for a potential genius from the lower class to stand less chance of success than a raging idiot from the upper class.

The rigid class structure had its value as a splitting device, but it was a grotesquely wasteful system, and it is not surprising that it eventually succumbed. Its ghost goes marching on, but it has largely been replaced today by a much more efficient meritocracy, in which each individual is theoretically able to find his optimum level. Once there he can consolidate his social identity by means of the various pseudo-tribal groupings.

This meritocratic system provides an exciting format, but there is another side to it. With excitement goes strain. An essential feature of a meritocracy is that, although it avoids waste of talent, it also opens up a clear channel from the very bottom to the very top of the enormous super-tribal community. If any small boy can, on his personal merits, eventually become the greatest of leaders, then for every one who succeeds there will be vast numbers of failures. These failures can no longer put the blame on the external forces of the wicked class system. They must place it firmly where it belongs, on their own personal shortcomings.

It seems, therefore, that any large-scale, lively, progressive super-tribe must inevitably contain a high proportion of intensely frustrated status-seekers. The dumb contentment of a rigid, stagnant society is replaced by the feverish longings and anxieties of a mobile, developing one. How do the struggling status-seekers react to this situation? The answer is that, if they cannot get to the top, they do their best to create the illusion of being less subordinate than they really are. To understand this, it will help at this point to take a sidelong glance at the world of insects.

Many kinds of insects are poisonous, and larger animals learn to avoid eating them. It is in the interests of these insects to show a warning flag of some kind. The typical wasp, for example, carries a conspicuous colour pattern of black and yellow bands on its body. This is so distinctive that it is easy for a predatory animal to remember it. After a few unfortunate experiences it quickly learns to avoid insects bearing this pattern. Other, unrelated, poisonous insect species may also carry a similar pattern. They become members of what has been called a 'warning club'.

The important point for us, in the present context, is that some *harmless* species of insects have taken advantage of this system by developing colour patterns similar to those of the poisonous members of the 'warning club'. Certain innocuous flies, for instance, display black and yellow bands on their bodies that mimic the colour patterns of the wasps. By becoming fake members of the 'warning club' they reap the benefits without having to possess any real poison. The killers dare not attack them, even though they would, in reality, make a pleasant meal.

We can use this insect example as a crude analogy to help us to understand what has happened to the human status-seeker. All we have to do is to substitute the possession of dominance for the possession of poison. Truly dominant individuals will display their high status in many visible ways. They will wave their dominance flags in the form of the clothes they wear, the houses they live in, the way they travel, talk, entertain and eat. By wearing the social badges of the 'dominance club' they make their senior status immediately obvious, both to subordinates and to one another, so that they do not have constantly to reassert their dominance in a more direct way. Like the poisonous insects, they do not have to keep on 'stinging' their enemies, they only have to wave the flag that says they could if they wanted to.

It follows, naturally enough, that harmless subordinates can join the 'dominance club' and enjoy its benefits if only they can display the same flags. If, like the black-and-yellow flies, they can mimic the black-and-yellow wasps, they can at least create the illusion of dominance.

Dominance mimicry has, in fact, become a major preoccupation of the super-tribal status-seekers, and it is important to examine it more closely. First, it is essential to make a clear distinction between a status symbol and a dominance mimic. A status symbol is an outward sign of the true level of social dominance you have attained. A dominance mimic is an outward sign of the level of dominance you would like to attain, but have not yet reached. In terms of material objects, a status symbol is something you can afford; a dominance mimic is something you cannot quite afford, but buy all the same. Dominance mimics therefore frequently involve making major sacrifices in other directions, whereas true status symbols do not.

Earlier societies, with their more rigid class structures, clearly did not give so much scope for dominance mimicry. As I have already pointed out, people were much more content to 'know their station'. But the upgrading urge is a powerful force, and there were always exceptions, no matter how rigid the class structure. The dominant individuals, seeing their position weakened by imitation, reacted harshly. They introduced strict regulations and even laws to curb the mimicry.

The various rules of costume give a good example. In England, the law of the Westminster parliament of 1363 was concerned chiefly with regulating the fashion of dress in the different social classes, so important had this subject become. In Renaissance Germany, a woman who dressed above her station was liable to have a heavy wooden collar locked around her neck. In India, strict rules were introduced relating the way you folded your turban to your particular caste. In the England of Henry VIII no woman whose husband could not afford to maintain a light horse for the king's service was allowed to wear velvet bonnets or golden chains. In America, in early New England, a woman was forbidden to wear a silk scarf unless her husband was worth a thousand dollars. The examples are endless.

Today, with the breakdown of the class structure, these laws have became severely curtailed. They are limited now to a few special categories such as medals, titles and regalia, which it is still illegal, or at least socially unacceptable, to adopt without the appropriate status. In general, however, the dominant individual is far less protected against the practices of dominance mimicry than he once was.

He has retaliated with ingenuity. Accepting the fact that lower-status individuals are determined to copy him, he has responded by making available cheap, mass-produced imitations of high-status goods. The bait is tempting and has been eagerly swallowed. An example will explain how the trap works.

High-status wife wears a diamond necklace. Low-status wife wears a bead necklace. Both necklaces are well made;

the beads are inexpensive, but they are gay and attractive and make no pretence to be anything other than what they are. Unfortunately, they have low status value, and the low-status wife wants something more. There is no law or social edict preventing her from wearing a diamond necklace. By working hard, saving every penny, and eventually spending more than she can afford, she may be able to acquire a necklace of small but real diamonds. If she takes this step, adorning her neck with a dominance mimic, she starts to become a threat to the high-status wife. The difference in their status displays becomes blurred. High-status husband therefore puts on the market necklaces of large, fake diamonds. They are inexpensive and superficially so attractive that the low-status wife abandons her struggle for real diamonds and settles for the fake ones instead. The trap is sprung. True dominance mimicry has been averted.

On the surface this is not apparent. The low-status wife, sporting her flashy fake necklace, *seems* to be mimicking her dominant rival, but this is an illusion. The point is that the fake necklace is too good to be true, when judged against her general way of life. It fools no one, and therefore fails to act as an aid in raising her status.

It is surprising that the trick works so well and so often, but it does. It has infiltrated many spheres of life and has not been without its repercussions. It has destroyed a great deal of genuine but overtly low-status art and craft. Native folk art has been replaced by cheap reproductions of the great masters; folk music has been replaced by the gramophone record; peasant craftsmanship has been replaced by mass-produced plastic imitations of more expensive goods.

Folklore societies have been rapidly formed to bewail and reverse this trend, but the damage has already been done. At best, all they can achieve is to act as folk-culture taxidermists. Once the status race was opened up from the bottom to the top of society, there was no turning back. If, as I suggested earlier, society is repeatedly going to rebel against the dreary uniformity of this 'new monotony', then it will do so by giving birth to new cultural patterns rather than by propping up old, dead ones.

For the really serious status-climber, however, there is no rebellion. Nor, for him, do the cheap fakes provide a satisfactory answer. He sees them for what they are, a clever side-track, a mere fantasy version of true dominance mimicry. For him, the dominance mimics must be genuine articles, and he must always go one step farther than he can afford, when purchasing them, in order to give the impression that he is slightly more socially dominant than he in fact is. Only then does he stand a chance of getting away with it.

For safety's sake, he tends to concentrate on areas where cheap fakes are out of the question. If he can afford a small motor car, he buys a medium-sized one; if he can afford a medium-sized one, he buys a large one; if he can afford a single large one, he buys a second car as a runabout; if large cars become too common, he buys a small, but wildly expensive foreign sports car; if large rear-lights become the fashion, he buys the latest model with even bigger ones, 'to let the people behind know he is in front', as the advertisers so succinctly express it. The one thing he does not do is to buy a row of life-sized, cardboard models of Rolls-Royces and display them outside his garage. There are no fake diamonds in the world of the status-climbing fanatic.

Motor cars are a single example, and an important one because they are so public, but the ardent status-struggler cannot stop there. He must extend himself and his bank balance in all directions if he is to paint a convincing picture for his higher-status rivals. The whole hire-purchase, mortgage and overdraft system depends for its survival on this expression of the powerful up-grading urge in terms of dominance mimics.

Unhappily, the extravagant trappings of the unrelenting status-seeker acquire such an importance that they appear to be more than they are. They are, after all, only mimics of dominance, not dominance itself. True dominance, true social status, is related to the possession of power and influence over super-tribal subordinates, not to the possession of a second colour television set. Of course, if you can easily afford a second colour television set, then it is a natural reflection of your status and acts as a true status symbol. A second colour television set, when you can only just afford the first one, is a different matter. It may help to impress on the members of the social level above you that you are ready to join them, but it in no way *ensures* that you will do so. All your rivals, at your own level, will be busily installing their own second colour television sets with the same idea in mind, but it is the fundamental law of the hierarchy that only a few from your level will make the grade to the one above. They, the lucky ones, can justifiably hang wreaths around their second colour television sets. Their dominance mimics worked the trick. All the rest, the power failures, must sit there, surrounded by the expensive clutter of the dominance mimics that have suddenly revealed themselves for what they are: illusions of grandeur. The realization that although they are valuable aids to successful dominance ladder-climbing, they do not actually guarantee it, is a bitter pill to swallow.

The damage caused by the exaggerated pursuit of dominance mimicry can be enormous. It not only leads to a condition of depressing disillusionment for the less successful status-seekers, it also demands such great efforts of the super-tribesman that he may have little time or energy for anything else.

The male status-seeker who indulges in an excess of dominance mimicry is often driven to neglect his family. This forces his mate to take over the masculine parental role in the home. Taking such a step provides a psychologically damaging atmosphere for the children, which can easily warp their own sexual identities when they mature. All that the young child will see is that its father has lost his leading role inside the family. The fact that he has sacrificed it to a struggle for dominance outside, in the larger sphere of the super-tribe, will mean little or nothing in the child's brain. If it matures with a well-balanced state of mental health it will be surprising. Even the older child, who comes to understand the super-tribal status race and boasts about his father's status achievements, will find them small compensation for the absence of an active paternal influence. Despite his mounting status in the outside world, the father can easily become a family joke.

It is very bewildering for our struggling super-tribesman. He has obeyed all the rules, but something has gone wrong. The super-status demands of the human zoo are cruel indeed. Either he fails and becomes disillusioned, or he succeeds and loses control of his family. Worse still, he can work so hard that he loses control of his family and *still* fails.

This brings us to another and more violent way in which certain members of the super-tribe can react to the frustrations of the dominance struggle. Students of animal behaviour refer to it as the re-direction of aggression. At the best of times it is an unpleasant phenomenon; at worst, it is literally lethal. One can see it very clearly when two rival animals meet. Each wants to attack the other and each is afraid to do so. If the aroused aggression cannot find an outlet against the frightening opponent who caused it, then it will find expression elsewhere. A scapegoat is sought, a milder, less intimidating individual, and the pent-up anger is vented in his direction. He has done nothing to warrant it. His only crime was to be weaker and less frightening than the original opponent.

In the status race it frequently occurs that a subordinate dare not express his anger openly towards a dominant. Too much is at stake. He has to re-direct it elsewhere. It may land on his unfortunate children, his wife or his dog. In former times, the flanks of his horse also suffered; today, it is the gearbox of his car. He may have the luxury of staff subordinates of his own that he can lash with his tongue. If he has inhibitions in all these directions there is always one person left: himself. He can give himself ulcers.

In extreme cases, when everything seems utterly hopeless, he can increase his self-inflicted aggression to the maximum: he can kill himself. (Zoo animals have been known to inflict serious mutilations on themselves, biting their flesh to the bone when unable to get at their enemies through the bars, but suicide seems to be a uniquely human activity.) Views concerning the main causes of suicide have differed widely, but hardly anyone denies that re-directed aggression is a major factor. One authority went so far as to claim that: 'Nobody kills himself unless he also wants to kill others or at least wishes some other person to die.' This is perhaps slightly over-stating the case. A man who kills himself because of the pain of an incurable disease scarcely falls into this category. It would be fanciful to suggest that he wants to kill the doctor who has failed to cure him. What he wants is release from pain. But re-direction of aggression does seem to account for a large number of cases. Here are some of the facts that support this idea.

There is a higher suicide rate in big towns and cities than in rural areas. In other words, where the status race is hottest, the suicide rate is highest. There are more male suicides than female suicides, but the females are catching up fast. In other words, the sex that is most involved in the status race has the highest suicide rate, and now that females are becoming increasingly emancipated and joining in the race more, they are sharing its hazards. There is a higher rate of suicide during periods of economic crisis. In other words, when the status race gets into difficulties at the top, there is an increase of re-directed aggression down the hierarchy, with disastrous results.

There is a lower rate of suicide during times of war. The suicide curves for the present century show two huge dips during the periods of the two world wars. In other words, why kill yourself if you can kill someone else? It is the inhibitions about killing the people who are dominating and frustrating the potential suicide that force him to re-direct his violence. He has the choice of killing a less daunting scapegoat, or himself. In peace-time, inhibitions about killing make him turn most often towards himself, but during war-time he is ordered to kill, and the suicide rate goes down.

The relationship between suicide and murder is a close one. To a certain extent they are two sides of the same coin. Countries with a high murder rate tend to have a low suicide rate, and vice versa. It is as if there is just so much intense aggression to be let loose, and if it does not take the one form it will take the other. Which way it goes will depend on how inhibited a particular community is about committing murder. If the inhibitions are weak, then the suicide rate goes down. It is similar to the war-time situation, where inhibitions against killing were actively and purposely reduced.

By and large, however, our modern super-tribes are remarkably heavily inhibited where acts of murder are concerned. It is difficult for the majority of us, who have never had to toss the murder/suicide coin, to appreciate the conflict, although in theory it seems biologically more unnatural to kill oneself rather than someone else. Despite this the figures go the other way. In Britain during recent times the yearly suicide figures have hovered around the 5,000 mark, while the yearly (detected) murders have kept below the 200 level. What is more, if we look at these murders, we find something unexpected. Most of us gain our ideas about murder from newspaper reports and detective novels, but newspapers and thriller-writers tend to concentrate on murders that will sell copies of papers and books. In reality the most common form of homicide is an unglamorous and squalid little family affair in which the victim is a close relative. There were 172 murders in Britain in 1967, and 81 of these were of this type. Furthermore, in 51 cases the murderer followed his act of homicide by committing suicide. Many of these latter cases are of the kind where a man, driven to turn his frustrated aggression on to himself, first kills his loved ones and then himself. Often, it appears that he cannot bear to leave them behind to suffer from the mess he has made, and so dispatches them first. Students of murder have discovered that an interesting change may then come over the killer. If he does not finish the job off and add his own corpse quickly to the rest, he is likely to experience such an enormous relief from tension that he suddenly finds he no longer wants to kill himself. Society dominated and frustrated him to the point where he was ready to take his own life, but now the slaying of his family consummates his revenge on society so effectively that his depression lifts and he feels released. This leaves him in a difficult situation. There are bodies lying about and all the signs that he has committed a multiple murder, when in fact in was only part of a desperate suicide. Such are the nightmare extremes of re-directed aggression.

Most of us, happily, do not reach such extremes. Our families may experience nothing more than our arrival home occasionally in a disgrunted mood. Many super-tribesmen can find an outlet by watching other people kill villains on television or at the cinema. It is significant that in strongly subordinated or suppressed communities, the local cinemas show a remarkably high proportion of films of violence. In fact, it can be argued that the thrills of fictional violence have an appeal that is directly proportional to the degree of dominance frustration that is being experienced in real life.

Since all the large super-tribes, by their very size, involve extensive dominance frustration, the prevalence of fictional violence is widespread. To prove the point it is only necessary to compare the international sales of books by authors of violent fiction with those of other writers. In a recent survey of the all-time best-sellers in the world of fiction, the name of one author who specializes in extremes of violence appeared seven times in the top twenty, with a total score of over 34 million copies sold. In the world of television the picture is much the same. A detailed analysis of transmissions in the New York area in 1954 revealed that there were no less than 6,800 aggressive incidents in a single week.

Clearly there is a powerful urge to watch other people being subjected to the most extreme forms of domination. Whether this acts as a valuable and harmless outlet for suppressed aggression is a hotly debated point. As with dominance mimicry, the cause of violence-watching is obvious, but the value is dubious. Reading about or watching an act of persecution does not alter the reader's or watcher's real-life situation. He may enjoy the experience of the fiction while he is involved in it, but when it is over and he re-emerges into the cold light of reality, he is still as dominated as he was before. The relief from tension is therefore only a temporary one, like scratching an insect bite. What is more, scratching a bite is likely to increase the inflammation. Repeated involvement in fictional mayhem tends to intensify the preoccupation with the whole phenomenon of violence. The best that can be said for it is that, while it is going on, the audience itself is not performing acts of violence.

The action of re-directing aggression has often been referred to as the '... and the office-boy kicked the cat' phenomenon. This implies that only the lowest members of a hierarchy will turn their blocked anger on to an animal. Unhappily for animals this is not the case, and animal protection societies have the figures to prove it. Cruelty to animals has provided a major outlet for re-directed aggression from the times of the earliest civilizations, right up to the present day, and it has certainly not been confined to the lowest levels in the social hierarchy. From the slaughters of the Roman amphitheatres, to the bear-baiting of the Middle Ages and the bull-fighting of modern times, the infliction of pain and death on animals has undeniably had a mass appeal for members of super-tribal communities. It is true that ever since our early ancestors turned to hunting as a method of survival, man has inflicted pain and death on other animal species, but the motives were different in prehistoric times. In the strict sense, there was no cruelty then, the definition of cruelty being 'taking delight in another's pain'.

In super-tribal times we have killed animals for four reasons: to obtain food, clothing and other materials; to exterminate pests and vermin; to further scientific knowledge; and to experience the pleasure of killing. The first and second of these two reasons we share with our early hunting ancestors, the third and fourth are novelties of the super-tribal condition. It is the fourth that concerns us here.

The others may, of course, involve elements of cruelty, but it is not their primary characteristic.

The history of deliberate cruelty to other species has taken a strange course. The early hunter had a kinship with animals. He respected them. So, rather naturally, did the early farming peoples. But the moment that urban populations began to develop, large groups of human beings became cut off from direct contact with animals, and the respect was lost. As civilizations grew, so did man's arrogance. He shut his eyes to the fact that he was just as much an animal as any other species. A great gulf appeared: now only *he* had a soul and other animals did not. They were no more than brute beasts put on earth for his pleasure. With the spreading influence of the Christian religion, animals were in for a rough passage. We need not go into the details, but it is worth noting that as late as the middle of the nineteenth century, Pope Pius IX refused permission for the opening of an animal protection office in Rome on the grounds that man owed duties to his fellow men, but none to the lower animals. Later in the same century a Jesuit lecturer wrote: 'Brute beasts, not having understanding and therefore not being persons, cannot have any rights ... We have, then, no duties of charity nor duties of any kind to the lower animals, as neither to sticks and stones.'

Many Christians were beginning to have doubts about this attitude, but it was not until Darwin's theory of evolution began to have a major impact on human thought that man and the animals came closer together again. The re-acceptance of man's affinity with animals, which had been so natural to the early hunters, led to a second era of respect. As a result, our attitude towards deliberate cruelty to animals has been changing rapidly during the past hundred years; but despite increasingly powerful disapproval, the phenomenon is still very much with us. Public displays are rare, but private savageries persist. We may respect animals today, but they are still our subordinates, and as such are highly vulnerable objects for the unloading of re-directed aggression.

Next to animals, children are the most vulnerable subordinates and, despite greater inhibitions here, they too are subjected to a great deal of re-directed violence. The viciousness with which animals, children and other helpless subordinates are subjected to persecution is a measure of the weight of the dominance pressures imposed on the persecutors.

Even in war, where killing is glorified, this mechanism can be seen in operation. Sergeants and other N.C.O.s frequently dominate their men with extreme ruthlessness, not merely to produce discipline, but also to arouse hatred, with the deliberate intention of seeing this hatred re-directed at the enemy in battle.

Looking back, we can see now how the unnaturally heavy weight of dominance from above, which is an inevitable characteristic of the super-tribal condition, has taken its toll. The abnormality of the situation for the human animal, who only a few thousand years ago was a simple tribal hunter, has produced patterns of behaviour which, by animal standards, are also abnormal: the exaggerated preoccupation with dominance mimicry, the excitement of watching acts of violence, the deliberate cruelty towards animals, children and other extreme subordinates, the acts of murder and, if all else fails, the acts of self-cruelty and self-destruction. Our super-tribesman, neglecting his family to drag himself one more rung up the social ladder, gloating over the brutalities in his books and films, kicking his dogs, beating his children, persecuting his underlings, torturing his victims, killing his enemies, giving himself stress diseases and blowing his brains out, is not a pretty sight. He has often boasted about being unique in the animal world, and on this score he certainly is.

It is true that other species also indulge in intense status struggles and that the attaining of dominance is often a time-consuming element in their social lives. In their natural habitats, however, wild animals never carry such behaviour to the extreme limits observable in the modern human condition. As I said at the outset, only in the cramped quarters of zoo cages do we find anything approaching the human state. If, in captivity, a group of animals is assembled which is too numerous for the species concerned, and they are packed too tightly together, then, with an inadequate cage environment, serious trouble will certainly develop. Persecutions, mutilations and killings will occur. Neuroses will appear. But even the least experienced zoo director would never contemplate crowding and cramping a group of animals to the extent that man has crowded and cramped himself in his modern cities and towns. *That* level of abnormal grouping, the director would predict with confidence, would cause a complete fragmentation and collapse of the normal social pattern of the animal species concerned. He would be astonished at the folly of suggesting that he should attempt such an arrangement with, say, his monkeys, his carnivores or his rodents. Yet mankind does this willingly to himself; he struggles under just these conditions and somehow manages to survive. By all the rules, the human zoo should be a screaming mad-house by now, disintegrating into complete social confusion. Cynics might argue that this is indeed the case, but plainly it is not. The trend towards denser living, far from abating, is ever gaining momentum. The various kinds of behaviour disorders I have outlined in this chapter are startling, not so much for their existence as for their rarity in relation to the population sizes involved. Remarkably few of the struggling super-tribesmen succumb to the extreme forms of action I have discussed. For every desperate status-seeker, home-wrecker, murderer, suicide, persecutor or ulcer-nurser, there are hundreds of men and women who not only survive, but thrive under the extraordinary conditions of the super-tribal assemblages. This, more than anything else, is a truly astonishing testimony to the enormous tenacity, resilience and ingenuity of our species.

CHAPTER THREE

SEX AND SUPER-SEX

WHEN you put a piece of food into your mouth it does not necessarily mean that you are hungry. When you take a drink it does not inevitably indicate that you are thirsty. In the human zoo eating and drinking have come to serve many functions. You may be nibbling peanuts to kill time, or you may be sucking sweets to soothe your nerves. Like a wine-taster, you may merely savour the flavour and then spit the liquid out, or you may down ten pints of beer to win a wager. Under certain circumstances you may be prepared to swallow a sheep's eyeball in order to maintain your social status.

In none of these cases is the nourishment of the body the true value of the activity. This multi-functional utilization of basic behaviour patterns is not unknown in the world of animals, but, in the human zoo, man's ingenious opportunism extends and intensifies the process. In theory, this should fall on the credit side of our super-tribal existence. There can, however, be drawbacks if we handle the process clumsily. If we eat too much to soothe our nerves, we become over-weight and unhealthy; if we drink too much of certain liquids we damage our livers or develop addictions; if we experiment too wildly with new tastes we get indigestion. These difficulties arise because we fail to separate non-nutritional feeding and drinking from their primary nutritional roles. We baulk at the ancient Roman habit of tickling the throat with a feather to make the stomach disgorge unwanted food, and the avoidance of swallowing practised by the wine-taster is no more than an isolated exception to the general rule. Nevertheless, with appropriate caution, we can indulge in multi-functional feeding and drinking to a considerable extent without coming to any serious harm.

Where sexual behaviour is concerned the situation is similar, but it is much more complicated and deserves our special attention. Here there has been an even greater failure to separate non-reproductive sexual activities from their primary reproductive functions. This has not, however, prevented the human zoo from converting sex into multifunctional super-sex, despite the fact that the results are sometimes disastrous for the human animals concerned. Man's opportunism knows no bounds and it is inconceivable that an activity so basic and so deeply rewarding should have escaped diversification. In fact, of all our activities, it has, regardless of the dangers, become functionally the most elaborate, with no fewer than ten major categories.

In order to clarify the picture, it will help if we examine the different functions of sexual behaviour one by one. It is important to realize at the outset that, although these functions are separate and distinct, and sometimes clash with one another, they are not all mutually exclusive. Any particular act of courtship or copulation may serve several functions simultaneously.

The ten functional categories are these:

1. *Procreation Sex*

There can be no argument that this is the most basic function of sexual behaviour. It has sometimes been mistakenly argued that it is the only natural and therefore proper role. Paradoxically, some of the religious groups that claim this do not practise what they preach, monks, nuns and many priests denying themselves the very activity which they hold to be so uniquely natural.

An important point to be added here is that when a population becomes seriously overcrowded, the value of the procreative function of sex becomes greatly reduced. Eventually it becomes a nuisance. Instead of being a fundamental mechanism of survival, it changes into a potential mechanism of destruction. This happens occasionally with such species as lemmings and voles which, when conditions are exceptionally lush, breed themselves up to such a density that their populations explode in chaos, with an enormous loss in lives. It is also happening to the human species at this very moment and the human animal may soon have to face the imposition of obtaining a breeding licence before being permitted to indulge in procreation.

This is not a matter that can be treated lightly and in recent years it has led to a great deal of agitated debate. It is worth looking at both sides of the argument, an exercise that has become increasingly rare as the protagonists have pushed one another into more and more extreme positions.

The basic question is: dare we tamper with the procreative process? Or, as the other side would put it: dare we *not* tamper with it? The arguments usually rage at a philosophical, ethical or religious level, but how do they appear when we view them biologically?

If a human group opposes efficient techniques for limiting procreativity, it gains two advantages. Firstly, it will breed more rapidly than the groups that do employ modern contraceptive devices. By gaining in numbers, it can hope eventually to swamp the others out of existence—a fact that cannot fail to appeal to its leaders, whether military or religious. Secondly, it will ensure that its basic social units—the family groups—are strong. A mated pair is not only a sexual unit, it is also a parental unit and the more parentally occupied it is, the more stable it becomes.

There are strong arguments, but so also are the opposing ones. Proponents of efficient contraception can point out that it is no longer a question of one group gaining on another. Over-population has become a world-wide problem and must be viewed as such. In this respect we are one vast, global lemming colony, and if the explosion comes it will affect us all. Indeed, it is already doing so.

As regards the family unit, it can be argued that contraception is not creating an unnatural situation, but merely recreating a natural one. Before medical care, hygiene and other security devices of modern living existed, the family unit may have produced large numbers of offspring, but it also lost a high proportion of them. All that contraception does, when applied in moderation, is to advance these losses to a point in time before the human egg has been fertilized.

If we do not pursue a world-wide policy of contraception, then some other unavoidable population limiting factor will step in. As a species we are rapidly reaching saturation point and if we fail to reduce our fecundity by voluntary means, the existing populations will suffer for it. If prevention is better than cure, then contraception is the obvious choice. It is difficult to see how anyone could argue that preventing someone from living is worse than curing someone of being alive. The individual human being is not a simple organism, to be squandered carelessly. It is a high-quality product requiring years of growth and development, and it needs all the protection it can get. Yet the opponents of contraception persist in their views. If they win, the swarms of non-contraceived offspring they are encouraging into the world may live to see the total collapse of the whole of human society.

2. *Pair-formation Sex*

The human animal is basically and biologically a pair-forming species. As the emotional relationship develops between a pair of potential mates it is aided and abetted by the sexual activities they share. The pair-formation function of sexual behaviour is so important for our species that nowhere outside the pairing phase do sexual activities regularly reach such a high intensity.

It is this function that causes so much trouble when it clashes with the various non-reproductive forms of sex. Even if Procreation Sex is successfully avoided and no fertilization takes place, a pair-bond may still automatically start to form where none is intended. It is because of this that casual copulations frequently create so many problems.

If a copulator has had his or her pair-forming mechanism damaged in some way during childhood, so that he or she is incapable of 'falling in love', or if there is a temporary and deliberate suppression of the pair-forming urge, then a casual copulation may succeed and be enjoyed without any later repercussions. But it takes two to copulate, and the partner in such an encounter may not be so lucky. If his or her pair-forming mechanism is more active, a one-sided pair-bond may start to form as a result of the emotional intensity of the sexual actions. The inevitable outcome of this is that society becomes littered with 'broken hearts', 'hang-ups' and 'abandoned lovers' who subsequently find it extremely difficult to form a new pair-bond with a fresh partner.

Only when the pair-bonding mechanism has been equally damaged or is equally suppressed in both partners can a casual human copulation be performed without undue risk. Even then, there is always the danger that the strength of the sexual response of one partner may be such that, for him or her, it will start to repair the bonding damage or disinhibit the bonding urge.

3. *Pair-maintenance Sex*

Once a pair-bond has been successfully formed, sexual activities still function to maintain and reinforce the bond. Although these activities may become more elaborate and *ex*tensive, they usually become less *in*tensive than those of the pair-forming stage, because the pair-forming function is no longer operating.

This distinction between the pair-forming and pair-maintaining functions of sexual activity is clearly illustrated whenever the members of a long-established mated pair are separated from one another for a period of time by war, business or some other external demand. When they are reunited there is typically a resurgence of high sexual intensity on the first nights they are together again, as they go through a minor re-bonding process.

There is one apparent contradiction that must be disposed of here. In some cultures, where the natural biological process of 'falling in love' is interfered with by arranged marriages or by anti-sexual propaganda, a young couple may find themselves newly married without even the beginnings of pair-bonding, or with a strongly inhibited approach to copulatory activity. In such cases they may report that (if they are lucky) their sexual behaviour becomes more intense at a later stage. For them, the pair-maintenance phase seems, at first sight, to be more sexually intense than the pair-formation stage, apparently reversing the correlation I have described. But this is not a real contradiction, it is simply that the true pair-forming stage has been artificially delayed.

Such couples are not always this fortunate. What frequently happens in such cases is that the family unit has to rely on external social pressures to hold it together, rather than the more basic, and more reliable, internal bonding process. If a marriage partner remains biologically 'unbonded' in this way, there is a considerable danger that a powerful extra-marital pair-bond will suddenly form. The true pair-forming capacity will be lying idle, so to speak, and will be all too ready to leap into action, causing havoc to the officially recognized 'pseudo-bond'.

There is a different kind of hazard for the young couple that *does* manage to base its marriage on the formation of a true pair-bond. This hazard is not caused by anti-sexual propaganda, but rather by a surfeit of pro-sexual propaganda, which can lead them to suppose that the very high intensity of the pair-formation stage should persist even after the pair has been fully formed. When it inevitably fails to do so, they imagine that something has gone wrong, whereas in reality they have merely reached the normal pair-maintenance sexual phase. The case for reproductive sex can be over-stated as well as under-stated, and either way may lead to trouble.

These first three categories—Procreation, Pair-formation and Pair-maintenance Sex—together make up the primary reproductive functions of human sexual behaviour. Before moving on to examine the non-reproductive patterns, there is one final, general comment that is relevant here. Individuals whose pair-bonding mechanism has run into some sort of trouble have occasionally found it convenient to argue that there is no such thing as a biological pairing urge in the human species. 'Romantic love', as they prefer to call it, is looked upon as a recent and highly artificial invention of modern living. Man, they argue, is basically promiscuous, like so many of his monkey relatives. The facts, however, are against this. It is true that, in many cultures, economic considerations have led to a gross distortion of the pair-forming pattern, but even where this pattern's interference with officially planned 'pseudo-bonds' has been most rigorously suppressed, with savage penalties and punishments, it has always shown signs of re-asserting itself. From ancient times, young lovers who have known that the law may demand no less then their lives if they are caught, have nevertheless found themselves driven to take the risk. Such is the power of this fundamental biological mechanism.

4. *Physiological Sex*

In the healthy adult human male and female there is a basic physiological requirement for repeated sexual consummation. Without such consummation, a physiological tension builds up and eventually the body demands relief. Any sexual act that involves an orgasm provides the orgasmic individual with this relief. Even if a copulation fails to fulfil any of the other nine functions of sexual behaviour, it can at least satisfy this basic physiological need. For an unmated and otherwise sexually unsuccessful male, a visit to a prostitute can serve this function. A more widespread solution, and one that is indulged in by both sexes, is masturbation.

A recent American study revealed that as many as 58 per cent of females and 92 per cent of males in that culture masturbate to orgasm at some time in their lives. Because this sexual act does not involve a partner and cannot therefore lead to fertilization, puritanical attempts have been made to suppress it at various times in the past, and all kinds of strange superstitions have grown up around it. The list of disaster that were supposed to threaten the masturbator included: desiccation, sterility, emaciation, frigidity, paroxysm, pallid complexion, hysteria, dizziness, jaundice, deformed figure, insanity, insomnia, exhaustion, pimples, pain, death, cancer, stomach ulcers, genital cancer, digestive upsets, headaches, appendicitis, weak hearts, kidney troubles, lack of hormones and blindness. This incredible collection of catastrophes would be amusing were it not for the untold miseries and fears the dire warnings must have caused, year after year and century after century. Happily these totally false superstitions are at last beginning to lose ground and a great deal of unnecessary anxiety is fading with them.

If no active sexual outlet is obtained, the body may take charge of the situation itself. Both male and female celibates are likely to undergo spontaneous orgasms while sleeping. Both sexes experience erotic dreams which may be accompanied by full orgasmic muscle responses and genital secretions in the female, and by 'nocturnal emissions' in the male.

Spontaneous orgasms appear to occur in even the most strictly abstemious and devoutly religious individuals, when they are described in rather different terms and referred to as religious frenzies, ecstacies or trances. St Theresa, for instance, described how a vision of an angel came to her: 'In his hands I saw a long golden spear and at the end of the iron tip I seemed to see a point of fire. With this he seemed to pierce my heart several times so that it penetrated to my entrails. When he drew it out I thought he was drawing them out with it and he left me completely afire with a great love of God. The pain was so sharp that it made me utter several sharp moans; and so excessive was the sweetness caused me by this intense pain that one can never wish to lose it.'

Unfortunately we know far too little about the spontaneous sexual outlets of extreme celibates to be able to make any firm statements concerning how widespread or how frequent these orgasms are. We do, however, know that individuals who have carried on an active sex life and are then confined in prison frequently show a marked increase in orgasmic dreaming. In one study, involving 208 female prisoners, this was found to be true for over 60 per cent of the group.

It would be wrong to give the impression, however, that orgasmic dreaming acts *solely* as a compensating device helping to keep up the sexual output when other more active outlets are missing. There is more to it than that, as there is, of course, with prostitution and masturbation, which serve other sexual functions as well. Some individuals, for example, show an increase in the frequency of orgasmic dreaming during periods when they are experiencing an unusually *high* frequency of active copulation, on the hypersensitizing principle of 'the more you get, the more you want'. However, this does not invalidate the evidence that spontaneous orgasm can and does occur as a response to sexual deprivation. It merely means that the phenomenon is more complex. But here we are only concerned with the simple, 'relief from physiological tension' function of sexual behaviour, and it is clear that this must be included as one of the ten basic functional categories of human sexual behaviour.

Physiological Sex can also be observed in other animal species and it is worth taking a look at a few examples. As one might expect, they are most readily encountered in the animal zoo, rather than in the wild state. Many zoo animals have been seen to masturbate when kept in isolation. This is most commonly observed in captive monkeys and apes. In males, the penis is stimulated sometimes by the hand or foot, sometimes by the mouth and sometimes by the tip of the prehensile tail. Male elephants stimulate their penises with their trunks and female elephants kept in a group without a male stimulate one another's genitals with their trunks. Even a male lion, kept in isolation in a zoo cage, has been seen to heave itself up into an inverted position against a wall and masturbate with its paws. Male porcupines have been observed to walk around on three legs, holding one forepaw on their genitals. One male dolphin developed the pattern of holding its erect penis in the powerful jet of the water intake of its pool. Sex-dreaming also seems to occur in animals and in domestic cats the erection of the penis while asleep has been observed to lead to full ejaculation.

5. *Exploratory Sex*

One of man's greatest qualities is his inventiveness. In all probability our monkey ancestors were already endowed with a reasonably high level of curiosity; it is a characteristic of the whole primate group. However, when our early human ancestors took to hunting, they undoubtedly had to develop and strengthen this quality and magnify their basic urge to explore all the details of their environment. It is clear that exploration became an end in itself, leading man on to fresh pastures and fresh achievements, always investigating, always asking new questions, never satisfied with old answers. So powerful did this urge become that it soon began to spread into all other areas of behaviour. With the arrival of the super-tribal condition, even simple patterns like locomotion were explored for possible variations. Instead of being satisfied with walking and running, we tried out hopping, skipping, leaping, marching, dancing, handstanding, vaulting, diving and swimming. Half the reward was in the experimentation itself, the actual discovering of a new variation. (Repeatedly indulging in it, following the discovery, was the second half of the reward, but we are not concerned with that for the moment.)

In the sexual sphere, this trend led to a wide range of variations on the sexual theme. Sexual partners began to experiment with new forms of mutual stimulation. Ancient sexual writings record in detail the great diversity of novel sexual movements, pressures, sounds, contacts, scents and copulatory positions that were the subject of erotic experimentation.

Although this was an inevitable development, paralleled by similar sensory explorations in other patterns, such as feeding behaviour, there were repeated attempts to suppress it in various cultures. The official reason given was often the one we have heard already; namely, that it represented an elaboration of sexual behaviour beyond what was necessary for the act of procreation. The significance of the development of exploratory sexual behaviour as an aid to the cementing of the pair-bond and the subsequent strengthening of the vital family unit was ignored. This was unfortunate, for one particularly important reason. As I have already mentioned, the intensity of love-making during the pair-forming stage wanes slightly after the pair-bond is fully formed. Theoretically, if the family unit is successful and remains unharassed by external forces, all should be well. It is an adaptive system because, if the exhausting intensiveness of the love-making of the young couple during pair-formation were prolonged indefinitely, it could well impair their efficiency in other activities. But the stresses and strains of the super-tribal condition *do* tend to harass the family unit. The external pressures are strong. The replacement of pair-forming intensiveness with exploratory extensiveness in later sexual activities is the ideal solution, and despite its repeated suppression it is still very much with us today.

There is only one drawback. The excitement of exploring

novel forms of sexual stimulation, when practised between the members of a mated pair, serves the family unit well. But it can take another form. The urge for novelty can be satisfied not only by exploring new patterns with a familiar partner, but also by exploring a new partner with familiar patterns, or, even more so, by exploring a new partner with new patterns.

The development of Exploratory Sex emerges, therefore, as a double-edged sword. Because our super-tribal cultures have laid increasing stress on the benefits of exploratory behaviour—our educational system, our great learning, our arts, sciences and technologies all depend on this—the exploratory urges in all our other patterns of behaviour have been similarly strengthened. In the sexual sphere this has frequently led to difficulties. The idea of a mated female attending practical classes in copulatory technique, or a mated male limbering up in a sexual gymnasium, is deeply offensive to their long-term sexual partners, since it interferes with the inherent exclusivity of the pair-bonding mechanism. Sexual experiments away from the mate therefore have to be made privately and in secret, and the new hazard of pair-bond betrayal enters the scene. The ancient and fundamental social nucleus of our species—the family unit—has suffered as a result, but somehow it has managed to survive.

These difficulties would not arise if we were a different kind of species, if we laid eggs in the sand like a turtle and left them to hatch out by themselves. But for us, with our heavy parental duties, sexual experiments outside the pair-bond have two dangers. They not only provoke powerful sexual jealousies, but they also encourage the accidental formation of new pair-bonds, to the long-lasting detriment of the offspring of the family units involved. Complex sexual combinations and communes may have worked from time to time, but unqualified successes seem to have been isolated rarities, limited to exceptional and unusual personalities. Only the most ruthless intellectual control by all parties concerned will permit sexual experiments of this kind to operate smoothly.

Even the rather widespread harem system, when viewed against the broader background of super-tribal success, has not fared well, and some scholars have pointed an accusing finger at it as an important factor in the social decline of the cultures concerned.

As with the other nine categories of sexual behaviour, the exploratory function is basic enough to be observable in other animal species. Since it requires a high level of inventiveness, it is not surprising that it is limited largely to the higher primates. The great apes, in particular, show a considerable range of sexual experiments when living under conditions of captivity, including a number of copulatory postures not seen in their wild counterparts.

6. *Self-rewarding Sex*

It is impossible to draw up a complete list of the functions of sex without including a category based on the idea that there is such a thing as 'sex for sex's sake'; sexual behaviour, the performance of which brings its own reward, regardless of any other considerations. This function is closely related to the last one, but they are nevertheless distinct.

The relationship between Exploratory Sex and Self-rewarding Sex is rather like the relationship between exploring and playing a game, or between random play and structured play in children. When children burst out into a new play environment, they usually start off with a great deal of erratic rushing about and investigating. As time goes by, this almost random behaviour settles down into a patterned sequence. A play-structure emerges and a 'game' is born. A particular environment may lend itself to a climbing game, or a hiding game, or a hunting game, and once such a game has been developed it may be repeated eagerly on later occasions without undue variation. If it proves to be a rewarding pattern it will be returned to over and over again, even though it is no longer a novelty. The initial, erratic behaviour was exciting because it was Exploratory Play; the later, repeated pattern is exciting as Self-rewarding Play.

The parallel with Exploratory Sex and Self-rewarding Sex is obvious enough. Many highly satisfying copulatory incidents occur between the members of a mated pair that are deliberately not aimed at procreation, that are far in excess of the demands of pair-maintenance and that do not involve the introduction of novel experiments. They therefore fall into the present functional category. They represent Self-rewarding Sex, or, if you prefer, pure eroticism. They are to the copulator what gastronomy is to the feeder, or what aesthetics is to the artist. It is inconsistent to sing the praises of exquisite gastronomic experiences, or of sublime aesthetic experiences, while at the same time condemning beautiful erotic experiences. Yet this has often been done. It is true that undue excess can sometimes create problems, but then so can undue excesses with gastronomy or aesthetics. Extreme cases of sexual athleticism can prove so exhausting that there is little energy left for anything else, and the pattern of living becomes unbalanced, just as extreme indulgence in feeding can cause serious obesity and loss of physical health, and extreme obsession with aesthetic problems can lead to a damaging disregard for other aspects of social life. The same basic rules apply in each case.

Preoccupation with action for action's sake implies the existence of some degree of spare time and spare energy.

This in turn implies that the basic survival needs are being taken care of. In humans this means an urban society. In animals it means life in a zoo, with food supplied and enemies eliminated, and it is there, not surprisingly, that we find the examples of animal hyper-sexuality.

7. *Occupational Sex*

This is sex operating as occupational therapy, or, if you prefer, as an anti-boredom device. It is closely related to the last category, but again can be clearly distinguished from it. There is a difference between having spare time and being bored. Self-rewarding Sex can occur as just one of many ways of constructively utilizing the spare time available, with not the slightest sign of any boredom syndrome on the horizon. The function is the positive pursuit of sensory rewards. Occupational Sex, by contrast, functions as a therapeutic remedy for the negative condition produced by a sterile and monotonous environment. Mild boredom produces listlessness and a lack of direction or motivation. Intense boredom, in a really bleak, empty environment, has a different impact. It creates anxiety and agitation, irritability and eventually anger.

Experiments with students who were placed alone in featureless cubicles, wearing opaque goggles and heavy gloves that made small hand actions impossible, produced startling results. As the hours went by, they became increasingly unable to relax. They went to extreme lengths to invent any kind of trivial action they could perform in the limited circumstances. They began to whistle, talk to themselves, tap out rhythms, anything at all to break the monotony, no matter how absurd the activity. After several days they suffered from signs of severe stress and found the conditions so unbearable that they could not continue.

Intense boredom is not, therefore, a matter of lying around doing nothing, it is precisely the opposite. A point is reached where *any* activity will do, just so long as some kind of behaviour output can be achieved. The situation is too threatening to enjoy the sensitive pleasures typical of self-rewarding activities; it is more a question of stopping the pain of gross inactivity. Being under-active is damaging to the nervous system and the brain does its utmost to protect itself.

Under normal conditions of boredom—that is to say an empty environment, but not as deliberately empty as the one in the student experiments—the object most readily available for breaking the monotony is the subject's own body. If there is nothing else, there is always that. Nails can be bitten, noses can be picked, hair can be scratched; and the body can always be provoked to produce a sexual response. Since the goal is to produce the maximum amount of stimulation, sexual activities in this situation often become brutal and painful and sometimes even lead to severe mutilation, or physical injury of the genitals. The pain they cause is, in a sense, a bizarre part of the therapy, rather than an accidental outcome of it. Savage and prolonged masturbation is typical of this phenomenon, perhaps involving tearing of the skin, or the insertion of sharp objects into the genital tracts.

Extreme forms of Occupational Sex can be observed in human prisoners who have been forcibly cut off from their normal, stimulating environments. This is not a matter of Physiological Sex—a much smaller amount of indulgence would satisfy the specific physiological demands.

The phenomenon can also be seen in the case of pathological introverts. Here it may occur in environments that appear, superficially, to be adequately stimulating. A closer examination soon reveals, however, that although the individuals concerned seem to be surrounded by exciting stimuli, they are cut off from them by their abnormal psychological condition. They are psychologically starving in the midst of plenty. If for some reason they have become intensely anti-social and mentally isolated, unable to make contact with the ordinary world around them, they may be suffering from under-stimulation just as intense as that experienced by the physical prisoners in their cells. For the extreme isolates, whether physical or mental, the painful excesses of Occupational Sex become a lesser evil than total, moribund inactivity.

Zoo animals kept in sterile cages exhibit similar responses. When isolated from their mates they may exhibit Physiological Sex. Free from the pressures of finding food and avoiding enemies, and with spare time on their hands, they may indulge in Self-rewarding Sex. But driven to extremes of boredom, they may resort to Occupational Sex of a drastic kind. Some male monkeys become obsessional masturbators. Male ungulates kept with females, but with nothing else to do, may literally worry their mates to death, harrying them and chasing them beyond all natural limits. Apes have been known to behave in the same way. One male orang-utan living in an empty cage, when provided with a female, mated with her and embraced her so persistently that she temporarily lost the use of her arms and had to be removed. Monkeys or apes that have been reared away from their own kind frequently find it impossible to adjust to social life when introduced as adults to a group of their own species. Like the mentally disturbed human being who 'lives in a world of his own', they may huddle in a corner and continue to indulge in solitary Occupational Sex while only a few feet away from a receptive mate. This is very common in zoo chimpanzees, which are all too often reared in isolation as pets and then thrown together as adults. One pair with abnormal childhoods, that were kept as a 'married couple' in a cage with no other companions, repeatedly engaged in a great deal of sexual behaviour, but it was never directed at one another. Although they shared the enclosure, they were both mentally isolated. Sitting apart from each other, they would both masturbate regularly in a variety of ways. The female used small branches, or pieces of wood that she tore off the walls with her teeth and inserted into her vagina, performing these actions while the male stimulated his penis in another corner.

8. *Tranquillizing Sex*

Just as the nervous system cannot tolerate gross inactivity, so it rebels against the strains of excessive overactivity. Tranquilizing Sex is the other side of the coin from Occupational Sex. Instead of being anti-boredom, it is anti-turmoil. When faced with an overdose of strange, conflicting, unfamiliar or frightening stimuli, the individual seeks escape in the performance of friendly old familiar patterns that serve to calm his shattered nerves. When the pressures of living are heavy, the stressed victim can tranquillize himself by resorting to actions that he knows will bring him the satisfaction of a consummatory reward. In his stressed, overactive state he is unable to push anything through to a conclusion. He is tugged this way and that, never able to resolve specific problems because of constant interferences and the confusion of blocked pathways. His frustrations mount until any simple familiar act, no matter how irrelevant to the major preoccupations, will provide a welcome release, if only it can be performed without obstruction.

Trivial actions such as smoking a cigarette, chewing gum or taking a drink, help to pacify the anxious. Tranquillizing Sex operates in the same way. The soldier at war, waiting for battle, or the business executive in the middle of a crisis, may seek momentary peace in the arms of a responsive female. The personal, emotional involvement can be at a minimum, the actions stereotyped. In a way, the more automatic it is the better, because his brain is already over-involved and seeks only simplicity.

This is similar to the animal activity known as 'displacement activity'. When two rival animals meet and come into conflict with one another, each wants to attack the other, but each is afraid to do so. Their behaviour is blocked and in their thwarted, frustrated condition they may turn aside to perform simple, irrelevant actions, such as grooming themselves, nibbling at food or fiddling with nest material. These displacement actions do not, of course, resolve the original conflict, but provide a momentary respite from the stressed condition. If a female happens to be near by she may be briefly mounted, and as in the human cases the action is usually stereotyped and simple.

9. *Commercial Sex*

Prostitution has already been mentioned, but only from the point of view of the customer. For the prostitute herself the function of copulation is different. Subsidiary factors may be operating, but primarily and overwhelmingly it is a straightforward commercial transaction. Commercial Sex of a kind also figures as an important function in many marriage situations, where a one-sided pair-bond exists: one partner simply provides a copulatory service for the other in exchange for money and shelter. The provider who has developed a true pair-bond has to accept a mock one in return. The woman (or man) who marries for money is, of course, functioning as a prostitute. The only difference is that whereas she, or he, receives indirect payment, the ordinary prostitute has to operate on a pay-as-you-lay basis. But whether the system is organized on long-term or short-term contracts, the function of the sexual behaviour involved is fundamentally the same.

A milder form of sex-for-material-gain is executed by strip-teasers, dance hostesses, beauty queens, club girls, dancers, models and many actresses. For payment, they provide ritualized performances of the earlier stages of the sexual sequence, but (in their official capacities) stop short of copulation itself. Compensating for the incompleteness of their sexual patterns, they frequently exaggerate and elaborate the preliminaries that they offer. Their sexual postures and movements, their sexual personality and anatomy, all tend to become magnified in an attempt to make up for the strict limitations of the sexual services they provide.

Commercial Sex appears to be rare in other species, even in zoos, but a form of 'prostitution' has been observed in certain primates. Female monkeys in captivity have been seen to offer themselves sexually to a male as a means of obtaining food morsels scattered on the ground, the sexual actions distracting the male from the business of competing for food.

10. *Status Sex*

With this, the final functional category of sexual behaviour, we enter a strange world, full of unexpected developments and ramifications. Status Sex infiltrates and pervades our lives in many hidden and unrecognized ways. Because of its complexity, I omitted it from the last chapter so that I could deal with it more fully here. It will help if we start by examining the form it assumes in other species, before we take a look at it in the human animal.

Status Sex is concerned with dominance, not with reproduction, and to understand how this link is forged we must consider the differing roles of the sexual female and the sexual male. Although a full expression of sexuality involves the active participation of both sexes, it is nevertheless true to say that, for the mammalian female, the sexual role is essentially a submissive one, and for the male it is essentially an aggressive one. (It is no accident of legal jargon that when a man sexually 'fondles' an unwilling female, his action is referred to as an indecent 'assault'.) This is not merely due to the fact that the male is physically stronger than the female. The relationship is an integral part of the nature of the copulatory act. It is the male mammal who has to mount the female. It is he who has to penetrate and invade his partner's body. An over-submissive female and an over-aggressive male are simply exaggerating their natural roles, but an aggressive female and a submissive male are completely reversing their roles.

The sexual action of a female monkey is to 'present' herself to the male by turning her rump towards him, raising it up conspicuously and lowering the head end of her body. The sexual action of the male monkey is to mount the female's back, insert his penis and make pelvic thrusts. Because, in a sexual encounter, the female submits herself and the male imposes himself, these actions have been 'borrowed' for use in primarily non-sexual situations requiring more general signals of submissiveness and aggressiveness. If female sexual 'presenting' signifies submissiveness, then it can be used in this way in a purely hostile encounter. A non-sexual female monkey can present her rump to a male as a sign that she is simply not aggressive. It acts as an appeasement gesture and functions as *an indication of her subordinate status*. In response, the male can mount her and make a few cursory pelvic thrusts, using these actions *purely to display his dominant status*.

Status Sex, used in this way, is a valuable device in the social lives of monkeys and apes. As a ritual of subordination and dominance, it avoids bloodshed. A male approaches a female aggressively, spoiling for a fight. Instead of screaming or attempting to flee, which would only feed the fire of his aggression, the female 'presents' herself to him sexually, the male responds, and they part, their relative dominance positions reaffirmed.

This is only the beginning. The value of Status Sex is such that it has spread to cover virtually all forms of aggressive encounter within the group. If a weak male is threatened by a strong one, the underling can protect himself by behaving as a pseudo-female. He signals his subordination by adopting the female sexual posture, offering his rump to the dominant male. The latter asserts his dominant status by mounting the weaker male, just as if he were dealing with a submissive female.

Precisely the same interaction can be observed between two females. An inferior female, threatened by a superior one, will 'present' to her and be mounted by her. Even juvenile monkeys will go through the same ritual, although they have not yet reached the adult sexual condition. This underlines the extent to which Status Sex has become divorced from its original sexual condition. The actions performed are still sexual actions, but they are no longer sexually motivated. Dominance has made a take-over bid for them.

The fact that sexual activities are being used repeatedly and frequently in this additional context explains the apparently orgiastic condition of some monkey colonies. Visitors to zoos often come away with the idea that monkeys are insatiable sexual athletes, ready, at the flick of a rump, to mate with anything, be it male or female, adult or juvenile. In one sense, of course, this is true. The observation is accurate enough. It is the interpretation that is wrong. Only when one understands the non-sexual motivation of Status Sex, does the picture become more balanced.

It may help to give an example from nearer home. Nearly everyone is familiar with the friendly, submissive greeting of a domestic cat, as it rubs the side of its body against a human leg, with its tail held stiffly upwards and the rear end of its body raised high. Both male and female cats do this and if, in response, we stroke their backs, we can feel them pushing the rear ends of their bodies up against the pressure of our hands. Most people accept this simply as a feline greeting gesture and do not question its origin or significance. In reality, it is another example of Status Sex. It is derived from the sexual presentation of the female cat towards the male, its original function being the pre-copulatory exposure of the vulva. But like the Status Sex presentation action of the monkeys and apes, it has now become emancipated from its purely sexual role and is performed by either sex when in a friendly, submissive condition. Because of the human cat-owner's size and strength, he is inevitably and permanently dominant as far as his pet is concerned. If contact is made after a temporary absence, the cat feels the need to re-establish its subordinate role, hence the greeting ceremony utilizing a submissive Status Sex display.

The feline pattern is a fairly simple one, but returning to the monkeys again, there are some striking anatomical extensions of Status Sex that we should examine before investigating the human condition. Many female monkeys possess bright red patches of swollen, naked skin in the rump region. These are conspicuously displayed to the male during the sexual rump-presentation action. They are also, of course, displayed when a female offers her rump submissively in Status Sex encounters. It has recently been pointed out that the males of some species have evolved mimics of these red patches on *their* rumps, as an enhancement of their submissive Status Sex displays towards more dominant individuals. For the females, the red rump patches serve a dual purpose, but for the males their function is *exclusively* concerned with Status Sex.

Switching from the submissive to the dominant Status Sex display, we can see a similar development. The dominant motion involves erection of the penis, and this too has been elaborated by the addition of conspicuous colours. In a number of species the males possess bright red penises, often surrounded by a vivid blue patch of skin over the scrotal region. This makes the male genitals as conspicuous as possible and males can frequently be seen sitting with their legs spread apart, displaying these bright colours to the maximum. In this way they can signal their high status without even moving. In some monkey species, males displaying in this manner sit at the edge of their group and, if another group comes near by, the red penis becomes fully erect and may be repeatedly raised to strike its owner's stomach. In ancient Egypt, the sacred baboon was seen as the embodiment of masculine sexuality. It was not only depicted in its Status Sex display posture in Egyptian paintings and carvings, but was even embalmed and buried in that posture, seventy days being spent on the embalming procedure and two days on the funeral ceremony. Obviously, the dominant Status Sex display of this species came through loud and clear, not only to other baboons, but also to ancient Egyptians. This was no accident, as we shall see in a moment.

Just as in some species the males have mimicked the submissive female displays, evolving their own red rump patches, so the females have, in some cases, mimicked the dominant displays of the males. Some female South American monkeys have evolved an elongated clitoris, which has virtually become a pseudo-penis. In certain species it is so similar in appearance to the true penis of the male, that it is hard to tell the sexes apart. This has given rise to a number of native legends in the areas where these animals live wild. Because they all appear to be male, the local populations believe them to be exclusively homosexual. (Strangely enough the female hyena has also evolved a similar pseudo-penis, but the myth that has emerged in Africa is that this species is hermaphrodite, each individual enjoying both male and female sexual activities.)

In a few species of monkeys the females have evolved a pseudo-scrotum as well as a pseudo-penis. As yet, we have little information about the way these fake male genitals are employed in the wild. We do know that certain male South American monkeys use penis erection as a direct threat to a subordinate. In the case of the little squirrel monkey, it has become the most important dominance signal in the animal's repertoire. It is more than merely sitting around with the legs apart. When in a threatening mood, a superior male of this species approaches close to an inferior and obtrusively erects his penis in the inferior's face. The pseudo-penis of the female monkeys does not, however, appear to be erectile; perhaps it is enough simply to display it as it is, towards an inferior monkey.

This, then, is the Status Sex situation in our closest relatives, the monkeys and apes. I have gone into it in some detail because it provides a useful evolutionary background against which to examine Status Sex developments in the human species. It makes some of the extraordinary lengths to which the human animal has gone, in this direction, a little easier to comprehend. Already, while reading the details of the monkey behaviour, you will, like the ancient Egyptians, have noticed certain similarities to the human condition. With men, as with monkeys, the submissive female sexual patterns and the dominant male patterns have come to stand for submissiveness and dominance in non-sexual contexts.

The ancient female pattern of presenting the rump to the male still survives as a gesture of subordination. Children are often forced to bend over for punishment in this posture. Also the buttocks are generally regarded as the most 'ridiculous' part of the human body, to be joked about and laughed at, or have pins stuck in them. Helpless victims in sado-masochistic pornography—not to mention popular comedy cartoon films and drawings—are frequently trapped with their buttocks in the air.

It is in the realm of dominant male patterns, however, that the human imagination has really run riot. The art and literature of civilization, since its earliest days, have been strewn with phallic symbols of all kinds. In recent times these have often become highly cryptic and far removed from their original source, the erect human penis, but it is still possible to observe more direct and overt phallic displays in the most primitive surviving cultures. Among New Guinea tribesmen, for example, the males make war wearing long tubes fitted on to their penises. These extensions, often well over a foot in length, are held in a near-vertical position by cords attached to the wearer's body. In other cultures, too, the penis is ornamented and artifically enlarged in a variety of ways.

Clearly, if the erection of the penis is used as a threatening display of male dominance, then it follows that the greater the erection, the greater the threat. The visual signals transmitting the intensity of the threat are of four kinds: as the penis becomes erect, it alters its angle, it changes from soft to hard, it increases in width and it grows in length. If all four of these qualities can be artificially exaggerated, then the impact of the display will be maximally enhanced. There is a limit to what can be done on the human body itself (which is more or less reached by the New Guinea tribesmen), but there is no such limit where human effigies are concerned. In drawings, paintings and sculptures of the human form the phallic display can be magnified at will. The average length of the erect penis in real life in $6\frac{1}{4}$ inches, which is less than one-tenth of the height of an adult male. In phallic statues, the length of the penis often exceeds the height of the figure.

Exaggerating the phallus still further, the depiction of a body is omitted altogether and the drawing or sculpture simply shows a huge, vertical, disembodied penis. Ancient sculptures of this kind, often rising many feet into the air, have been found in numerous parts of the world. Giant phallic statues nearly two hundred feet high guarded the temple of Venus at Hierapolis, but even these were exceeded in size by another ancient phallus that reputedly towered to the height of three hundred and sixty feet, or roughly seven hundred times the length of the physical organ it represented. It is said to have been covered in pure gold.

From obvious representations of this kind, it is only one more step to the world of phallic symbolism, where almost any long, stiff, erect object can take on a phallic role. We know from the dream studies of the psycho-analysts just how varied these symbols can be. But they are not confined to dreams. They are deliberately used by advertisers, artists and writers. They appear in films, plays and almost all forms of entertainment. Even when they are not consciously understood they can still make their impact, because of the very basic signal they transmit. They include everything from candles, bananas, neckties, broom-handles, eels, walking-sticks, snakes, carrots, arrows, water-hoses and fireworks, to obelisks, trees, whales, lamp-posts, skyscrapers, flag-poles, cannons, factory chimneys, space rockets, lighthouses and towers. These all have symbolic value because of their general shape, but in some cases a more specific property is involved. Fish have become phallic symbols because of their texture as well as their shape, and because they thrust themselves through the water; elephants because of their erectile trunks; rhinos because of their horns; birds because they rise up against gravity; magic wands because they give special powers to the magician; swords, spears and lances because they penetrate into the body; champagne bottles because they ejaculate when opened; keys because they are inserted into keyholes; and cigars because they are tumescent cigarettes. The list is almost endless, the scope for imaginative symbolic equations enormous.

All these images have been used, and in most cases used frequently, as objects representing masculinity. The tough, dominant male (or would-be tough, dominant male) who chews on his fat cigar and thrusts it into his companion's face, is fundamentally performing the same Status Sex display as the little squirrel monkey that spreads its legs and thrusts its erect penis into the face of a subordinate. Social taboos have forced us to employ cryptic substitutes for our aggressive sexual displays, but the imagination of man being what it is, this has not reduced the phenomenon, it has only diversified and elaborated it. As I explained in the chapter dealing with status and super-status, we have good reason in our super-tribal condition to make great play with our status devices, and this is precisely what we do in the case of Status Sex.

We can find examples of various kinds of improvements in phallic symbols taking place almost as we watch. The design of sports cars illustrates this well. They have always radiated bold, aggressive masculinity and have been considerably aided in this by their phallic qualities. Like a baboon's penis, they stick out in front, they are long, smooth and shiny, they thrust forward with great vigour and they are frequently bright red in colour. A man sitting in his open sports car is like a piece of highly stylized phallic sculpture. His body has disappeared and all that can be seen are a tiny head and hands surmounting a long, glistening penis. (The argument may be used that the shape of sports cars is controlled purely by the technical demands of streamlining, but the crowded traffic conditions of modern driving and the increasingly strict speed limits render this explanation nonsensical.) Even ordinary motor cars have their phallic qualities, and this may explain to some extent why male drivers become so aggressive and so eager to overtake one another, despite considerable risks and despite the fact that they all meet up again at the next set of traffic-lights, or, at best, only cut a few seconds off their journey.

Another illustration comes from the world of popular music, where the guitar has recently undergone a change of sex. The old-fashioned guitar, with a curvaceous, waisted body, was symbolically essentially female. It was held close to the chest, its strings loving caressed. But times have changed and its femininity has vanished. Ever since groups of masculine 'sex idols' have taken to playing electric guitars, the designers of these instruments have been at pains to improve their masculine, phallic qualities. The body of the guitar (now its symbolic testicles) has become smaller, less waisted and more brightly coloured, making it possible for the neck (its new symbolic penis) to become longer. The players themselves have helped by wearing the guitars lower and lower until they are now centred on the genital region. The angle at which they are played has also been altered, the neck being held in an increasingly erect position. With the combination of these modifications, the modern pop-group can stand on stage and go through the movements of masturbating their giant electric phalluses while they dominate their devoted 'slaves' in the audience. (The singer has to make do with fondling a phallic microphone.)

Contrasting with these phallic 'improvements' are a number of cases where phallic symbols have gone into decline or eclipse. As early civilizations (which, as I have said, were much more open in their use of phallic symbolism) were superseded, their obvious images were often cloaked and distorted. Perhaps the most startling example of this is the Christian cross. In earlier days this was a straightforward phallic symbol, the vertical piece representing the penis and the side pieces the testicles. It sometimes appears in a more explicit form in ancient, pre-Christian images, with a man's head at the top of the upright, his body being completely replaced by the stylized representation of his sexual organs in the form of the cross. It has been pointed out by one author that the acceptance of this symbol in a new role by Christians was probably facilitated by its earlier significance as a symbol of the 'life force'.

Another cross that has long since lost its original meaning is the famous Maltese cross. The ancient prehistoric ruins of Malta were full of phalluses, most of which have been lost, stolen or destroyed. Among them was a cross consisting of four huge stone phalluses which, as one writer put it, 'were subsequently metamorphosed by the virtuous Knights of St John' and served as their arms.

Easter festivities have also shown a reduction in phallic overtness. In ancient cultures this was often a time for baking phallic cakes. They were made in the shape of both male and female genitals, but today they have been transformed, appearing in certain countries as sweetmeats made in the shape of a fish (the male cake) and a doll (the female cake). The phallic nature of the fish symbol was also originally involved in the ritual of eating fish on Fridays, but this has long ago lost its sexual significance.

There are many other examples one could give. The bonfire, for instance, although still retaining an almost magical ritual quality on certain occasions, has lost its sexual properties. It was originally lit in a special way by rubbing a 'male' stick against a 'female' stick in an act of symbolic copulation, until a spark was generated and the bonfire burst into sexual flames.

Many buildings used to display carved phalluses on their outer walls to protect them against the 'evil eye' and other imagined dangers. These symbols, being aggressive, dominant Status Sex threats to the outside world, guarded the buildings and their occupants. In certain Mediterranean countries today one can still see symbols of this kind, but they have become less overtly sexual. They now usually consist of a pair of bull's horns, firmly fixed high up on an outer surface or the corner of a roof. However, despite these expurgations and censorships, which have turned the tree of carnal knowledge into the simple tree of knowledge and have replaced the obvious codpiece with the less obvious necktie, there are still areas where aggressive phallic symbols retain their original overt properties. In the realm of *insults* we find them still very much in evidence.

Verbal insults frequently take a phallic form. Almost all the really vicious swearwords we can use to hurl abuse at someone are sexual words. Their literal meanings relate to copulation or to various parts of the genital anatomy, but they are used predominantly in moments of extreme aggression. This again is typical of Status Sex and demonstrates very clearly the way in which sex is borrowed for use in a dominance context.

Visual insults follow the same trend, several kinds of

phallic actions being employed as hostile gestures. Sticking out the tongue originated in this way, with the protruded tongue symbolizing the erect penis. Hostile gestures known as giving the 'phallic hand' have existed in various forms for at least two thousand years. One of the most ancient consists of aiming the middle (i.e. second) finger, stiffly and fully extended, at the person being subjected to contempt. The rest of the fist is clenched. Symbolically, the middle finger represents the penis, the clenched thumb and first finger together represent one testicle, and the clenched third and fourth fingers together represent the other testicle. This gesture was popular in Roman times, when the middle finger was known as the *digitus impudicus*, or *digitus infamis*. It has become modified over the centuries, but can still be seen in many parts of the world. Sometimes the first finger is used instead of the middle finger, probably because this is a slightly easier posture to hold. Sometimes the first and second finger are extended together, emphasizing the size of the symbolic penis. It is usual today for this type of phallic hand to be jerked upwards into the air one or more times, in the direction of the insulted person, symbolizing the action of pelvic thrusting. The two extended fingers may be held together or separated into a V shape.

An interesting corruption of this last form appeared in recent times as the victory-V sign, which did far more than merely copy the first letter of the word victory. Its phallic properties also helped it. It differed from the insult-V by the position of the hand. In the insult-V the palm of the hand is held towards the insulter's face; in the victory-V it is held towards the admiring crowd of onlookers. This means, in effect, that the dominant individual making the victory-V sign is really making the insult-V, but on *their* behalf; for them, not against them. What they see as they watch their signalling leader is the same hand position as they themselves would see if *they* were making the insult-V. By the simple device of rotating the hand, the phallic insult becomes a phallic protection. As we have already observed, threat and protection are two of the most vital aspects of dominance. If a dominant individual performs a threat towards a member of his group, it insults the latter, but if the dominant one performs the same threat *away* from the group towards an enemy, or an imagined enemy, then his subordinates will cheer him for the protective role he is playing. It is astonishing to think that a leader can change his whole image simply by twisting his hand through 180 degrees, but such are the refinements of modern Status Sex signalling.

Another ancient form of 'phallic hand', also dating back at least two thousand years, is the so-called 'fig'. In this, the whole fist is clenched, but as it is aimed at the insulted person the thumb is pushed through between the base of the curled first and second fingers. The tip of the thumb then protrudes slightly, like the head of a penis, pointing at the subordinate or enemy. This gesture has spread throughout much of the world and almost everywhere is known as 'making the fig'. In English, the phrase 'I don't give a fig for him' means that he is not even worthy of an insult.

Many examples of these phallic hands have been found on ancient amulets and other ornaments. They were worn as protections against the 'evil eye'. Some people today might look upon such emblems as improper or obscene, but this was not the role they played when they were worn. They were used then, quite properly, as protective Status Sex symbols. In specific contexts the symbolic phallus was seen as something to be acclaimed and even worshipped as a magical guardian ready to ravage, not the members of the group, but threats coming from outside it. At the Roman festival of Liberalia an enormous phallus was carried in procession on a magnificent chariot into the middle of the public place of the town where, with great ceremony, the females, including even the most respectable matrons, hung garlands around it to 'remove enchantments from the land'. In the Middle Ages many churches had phalluses on their outer walls to protect them from evil influences, but in almost all cases these were later destroyed as 'depraved'.

Even plants were called into phallic service. The mandrake, a plant with phallus-shaped roots, was widely used as a protective amulet. It was improved for its symbolic role by embedding grains of millet or barley in it in the appropriate area, re-burying it for twenty days so that the grains sprouted, then digging it up again and trimming the sprouts to make them look like pubic hair. Kept in this form it was said to be so effective in dominating outside forces that it would double its owner's money each year.

It would be possible to go on and fill a whole book with examples of phallic symbolism, but the few I have selected are sufficient, I think, to show how widespread and varied this phenomenon is. We arrived at this topic by singling out just one of the elements of the aggressive male Status Sex display, namely penis erection. Other important developments have also occurred, however, which we should not overlook. The original, straightforward pattern of copulation is, for the male, as I have already stressed, a fundamentally assertive and aggressive act of penetration. Under certain conditions it can therefore function as a Status Sex device. A male can copulate with a female primarily to boost his masculine ego, rather than to achieve any of the other nine sexual goals I have listed in this chapter. In such cases he may speak of making a 'conquest', as if he has been fighting a battle rather than making love. And when I say he *speaks* of it I mean this literally, for boasting to other males is an important part of the Status Sex victory. If he keeps quiet about it, it can always feed his ego privately, but it provides him with a much stronger status boost if he tells his friends. Any female who finds out about this can be reasonably sure what kind of copulation she has been involved with. The details of pair-forming copulations are, by contrast, strictly private.

The male who uses females for Status Sex purposes is more concerned, in fact, with showing them off than with anything else. He may even be content to display his dependent females to his group without bothering to copulate with them. Providing they are clearly seen to be subordinate to him, this will often suffice.

The huge harems that were assembled by the rulers of certain cultures acted largely as a Status Sex device. They did not indicate the existence of multiple pair-bonding. Frequently a favourite wife emerged from the group of females, with whom some sort of pair-bond developed, but the business of Status Sex really came to dominate the whole scene. There was a simple equation: power = number of females in harem. Sometimes there were so many that the ruler had neither the time nor the energy to copulate with all of them, but as a symbol of virility he did attempt to sire as many offspring as possible. The would-be present-day harem overlord usually has to make do with a long series of females, dominating them one at a time, instead of gathering them around him all at once. He has to rely on his verbal reputation rather than on a massive visual display of sexual power.

It is relevant here to mention the special attitude of heterosexual Status Sex devotees towards homosexual males. It is an attitude of increased hostility and contempt, caused by the unconscious realization that 'if they won't join the game, they can't be beaten'. In other words, the homosexual male's lack of sexual interest in females gives him an unfair advantage in the Status Sex battle, for no matter how many females the heterosexual expert subdues, the homosexual will fail to be impressed. It then becomes necessary to defeat him by ridicule. Inside the homosexual world there will, of course, be Status Sex competition as vigorous as that found in the heterosexual sphere, but this in no way improves the understanding between the two groups, since the objects competed for are so different in the two cases.

If the modern Status Sex practitioner is unable to achieve real conquests, there are still a number of alternatives available to him. A mildly insecure male can express himself by telling dirty jokes. These carry the implication that he is aggressively sexual, but an obsessive, persistent dirty-joke-teller begins to arouse suspicions in his companions. They detect a compensation mechanism.

Males with a greater inferiority problem can frequent prostitutes. I have already mentioned other functions of this sexual activity, but status-boosting is perhaps the most important. The essential property of this form of Status Sex is that the female is being degraded. The male, providing he has a small amount of cash, can *demand* sexual submission. The fact that he knows the girl does not welcome his advances, but submits to them anyway, can even help to increase his feeling of power over her. Another alternative is the strip-tease display. The female, again for a small sum of money, has to strip herself naked in front of him, debasing herself and thereby raising the relative status of the watching males.

There is a savage satirical drawing on the subject of strip-tease, captioned simply 'tripes-tease'. It shows a naked girl who, having removed all her clothes and still being faced with shouts for 'more', makes an incision in her belly, and, with a seductive smile, starts to pull out her entrails to the beat of the music. This brutal comment reveals that with the subject of strip-tease we are moving into the realm of that extreme form of Status Sex expression, the realm of sadism.

It is an unpalatable but obvious fact that the more drastic the need for male ego-boosting, the more desperate the measures; the more degrading and violent the act, the greater the boost. For the vast majority of males, these extreme measures are unnecessary. The level of self-assertion achieved in ordinary social life is sufficiently rewarding. But under the heavy status pressures of super-tribal living, where there can be so few dominants and must be so many suppressed subordinates, sadistic thoughts nevertheless tend to proliferate. For most men they remain nothing more than thoughts; sadistic fantasies that never see the light of day. Some individuals go further, avidly studying the details of whipping, beatings and tortures in sadistic books, pictures and films. A few attend pseudo-sadistic exhibitions, and a very, very few become actual practising sadists. It is true that many men may be mildly brutal in their love-play and that some perform mock-sadist rituals with their mates, but the full-blooded sadist is fortunately a rare beast.

One of the most common forms of sadism is rape. Perhaps the reason for this is that it is so exclusive an act of the male that it expresses aggressive masculinity better than other types of sadistic activity. (Males can torture females and females can torture males. Males can rape females, but females cannot rape males.) In addition to the total domination and degradation of the female, one of the bizarre satisfactions of rape for the sadist is that the writhings and facial expressions of pain he produces in the female are somewhat similar to the writhings and facial expressions of a female experiencing an intense orgasm. Furthermore, if he then kills his victim, her immediately limp and passive condition presents a gruesome mimic of post-orgasmic collapse and relaxation.

An alternative pattern for meeker males is what might be described as 'visual rape'. Usually referred to as exhibitionism, this consists of suddenly exposing the genitals to a strange female, or females. No attempt is made to effect physical contact. The aim is to produce shame and confusion on the part of the unwilling female spectators, by presenting them with the most basic form of Status Sex threat display. Here we are right back to the penis threat of the little squirrel monkey.

Perhaps the most extreme kind of sadism is the torture, rape and murder of a small child by an adult male. Sadists of this type must suffer from feelings of the most intense status inferiority known to man. In order to obtain their ego boost, they are forced to select the weakest and most helpless individuals in their society and impose upon them the most violent form of domination they can perform. Happily, these extreme measures are rarely taken. They appear to be more common than they in fact are, because of the enormous publicity that such cases receive, but in reality they comprise only a minute fraction of the total picture of 'crimes of violence'. All the same, any super-tribe that contains even only a few individuals that are driven to dominance excesses of this kind must constitute a society operating under immense status pressures.

One final point on Status Sex: it is intriguing to discover that certain individuals with a demonstrably vast lust for power suffered from physical sexual abnormalities. The autopsy on Hitler, for instance, revealed that he had only one testicle. The autopsy on Napoleon noted the 'atrophied proportions' of his genitals. Both had unusual sex lives, and one is left to wonder just how much the course of European history would have been changed had they been sexually normal. Being structurally sexually inferior, they were perhaps driven back on to more direct forms of aggressive expression. But no matter how extreme their domination became, their urge for super-status could never be satiated, because no matter how much they achieved it could never give them the perfect genitals of the typical dominant male. This is Status Sex going full circle. First, the dominant male sexual condition is borrowed as an expression of dominant aggression. Then it becomes so important in this role that if there is something wrong with the sexual equipment, it becomes necessary to compensate by putting a stronger emphasis on pure aggression.

Perhaps there is something to be said for Status Sex (in its very mildest forms) after all. In its more ritualized and symbolic varieties it does at least provide comparatively harmless outlets for otherwise potentially damaging aggressions. When a dominant monkey mounts a subordinate, it manages to assert itself without recourse to sinking its teeth into the weaker animal's body. Swapping sex jokes in a bar causes less injury than having a punch-up or a brawl. Making an obscene gesture at someone does not give him a black eye. Status Sex has, in fact, evolved as a bloodless substitute for the bloody violence of direct domination and aggression. It is only in our overgrown super-tribes, where the status ladder stretches right up into the clouds and the pressures of maintaining or improving a position in the social hierarchy have become so immense, that Status Sex has got out of hand and gone to lengths that are as bloody as pure aggression itself. This is yet another of the prices that the super-tribesman has to pay for the great achievements of his super-tribal world and the excitements of living in it.

In surveying these ten basic functions of sexual behaviour we have seen very clearly the way in which, for the modern urban human animal, sex has become super-sex.

Although he shares these ten functions with other animals, he has pushed most of them much farther than the other species have ever done. Even in the most puritanical cultures, sex has played a major role, if only because there it was constantly on people's minds as something that needed suppressing. It is probably true to say that no one is quite so obsessed by sex as a fanatical puritan.

The influences at work in the trend towards super-sex have been interwoven with one another. The main factor was the evolution of a giant brain. On the one hand this led to a prolonged childhood and this in turn meant a long-term family unit. A pair-bond had to be forged and maintained. Pair-formation Sex and Pair-maintenance Sex were added to primary Procreation Sex. If active sexual outlets were not available, the ingenuity of the giant brain made it possible for various techniques to be employed to obtain relief from physiological sexual tension. Man's strengthened urge for novelty, his heightened curiosity and inventiveness, gave rise to a massive increase in Exploratory Sex. Because of its efficiency, the giant brain organized his life in such a way that man had more and more time to spare and a greater sensitivity available to him while filling it. Self-rewarding Sex, sex for sex's sake, was able to blossom. If there was too much time to spare, then Occupational Sex could step in. If, by contrast, the increased strain of super-tribal pressures and stresses became too heavy, then there was always Tranquillizing Sex. The added complexities of super-tribal life brought increased division of labour and trading, and sexual activity became involved here too, in the form of Commercial Sex. Finally, with the vastly magnified dominance and status problems of the huge super-tribal structure, sex was increasingly borrowed for use in a non-sexual context, as all-pervading Status Sex.

The greatest sexual complication to arise has been the

clash between the primarily reproductive categories (Procreation, Pair-formation and Pair-maintenance Sex), on the one hand, and the primarily non-reproductive categories, on the other. In the pre-pill days, when contraception was forbidden, rare or inefficient, Procreation Sex provided a major hazard for Exploratory Sex, Self-rewarding Sex and the rest. Even in the so-called 'post-pill paradise', which some have seen as heralding an epoch of wild promiscuity, the problem is far from solved, because of the persistence of the fundamental pair-bonding properties of human sexual encounters. Widespread, trouble-free promiscuity is a myth and always will be. It is a myth born of the wishful thinking of Status Sex, but it will for ever remain a wishful thought. Man's strong pair-forming urge stemming, in evolutionary terms, from his greatly increased parental duties, will persist no matter what technical advances are achieved with perfected contraception in the years to come. This does not mean that such advances will have no impact on our sexual activities. On the contrary, they will profoundly alter our behaviour. The triple pressure of improved contraception, dwindling venereal disease and ever swelling human population will together work towards a dramatic increase in non-reproductive forms of copulatory indulgence. There can be no doubt of this. Equally there can be no doubt that this will intensify the clash between these forms of sex and the demands of the pair-bond. Unhappily, as a result, the children will suffer along with their sexually confused parents.

It would be much easier if, like our monkey relatives, we had a lighter parental burden and were more truly biologically promiscuous. Then we could extend and intensify our sexual activities with the same facility that we magnify our body-cleaning behaviour. Just as we harmlessly spend hours in the bathroom, visit masseurs, beauty parlours, hairdressers, Turkish baths, swimming pools, sauna baths or Oriental bath-houses, so we could indulge in lengthy erotic escapades with anyone, at any time, without the slightest repercussions. As it is, it seems as if our basic animal nature will always stand in the way of this development, or will at least discourage it until such time as we have undergone some radical genetical change.

The only hope is that, as the clashing demands of super-sex grow more intense, we shall learn to play the game more deftly. It is, after all, possible to indulge oneself gastronomically without growing either fat or sick. With sex the trick is more difficult to accomplish, and society is littered with the bitter jealousies, forlorn heartbreaks, miserable, shattered families and unwanted offspring to prove it.

No wonder super-sex has become such a problem for the Urban Super-ape. No wonder it has so often been abused. It is capable of providing man with his most intense physical and emotional rewards. When it goes wrong, it is also capable of causing him his greatest miseries. As he has expanded it, elaborated it and manipulated it, he has magnified its potentials both as a reward and a punishment. But sadly there is nothing unusual in this. In many departments of human behaviour we find the same development. Even in medical care, for example, where the rewards are so obvious, the punishments are still there: it can so easily contribute to overcrowding which, in turn, leads to the proliferation of new stress diseases. It can also lead to hypersensitivity to pain. A New Guinea tribesman can have a spear removed from his thigh with more aplomb than a super-tribesman having a small splinter removed from his finger. But this is no reason for wanting to turn back. If our increased sensitivities can work both ways, we must make sure that they work the right way. The big change is that matters are now in our hands, or rather our brains. The tight-rope of survival which has been set up, and on which our species performs its daring tricks, has been raised higher and higher. The dangers have become greater, but then so have the thrills. The only snag is that when the tribes became super-tribes, someone took away our biological safety-net. It is up to us now to make sure that we do not crash to our deaths. We have taken over evolution and have no one to blame but ourselves. The strength of our animal properties is still carried securely within us, but so are our animal weaknesses. The better we understand them and the enormous challenges they are facing in the unnatural world of the human zoo, the better our chances of success.

CHAPTER FOUR

IN-GROUPS AND OUT-GROUPS

QUESTION: What is the difference between black natives slicing up a white missionary, and a white mob lynching a helpless Negro? Answer: very little—and, for the victims, none at all. Whatever the reasons, whatever the excuses, whatever the motives, the basic behaviour mechanism is the same. They are both cases of members of the in-group attacking members of the out-group.

In plunging into this subject we are entering an area where it is difficult for us to maintain our objectivity. The reason is obvious enough: we are, each one of us, a member of some particular in-group, and it is difficult for us to view the problems of inter-group conflict without, however unconsciously, taking sides. Somehow, until I have finished writing and you have finished reading this chapter, we must try to step outside our groups and gaze down on the battlefields of the human animal with the unbiased eyes of a hovering Martian. It is not going to be easy, and I must make it clear at the outset that nothing I say should be construed as implying that I am favouring one group as against another, or suggesting that one group is inevitably superior to another.

Using a harsh evolutionary argument, it might be suggested that if two human groups clash and one exterminates the other, the winner is biologically more successful than the loser. But if we view the species as a whole this argument no longer applies. It is a small view. The bigger view is that if they had contrived to live competitively but peacefully alongside one another, the species as a whole would be that much more successful.

It is this large view that we must try to take. If it seems an obvious one, then we have some rather difficult explaining to do. We are not a mass-spawning species like certain kinds of fish that produce thousands of young in one go, most of which are doomed to be wasted and only a few survive. We are not quantity breeders, we are quality breeders, producing few offspring, lavishing more care and attention on them and looking after them for a longer period than any other animal. After devoting nearly two decades of parental energy to them it is, apart from anything else, grotesquely *inefficient* to send them off to be knifed, shot, burned and bombed by the offspring of other men. Yet, in little more than a single century (from 1820 to 1945), no less than 59 million human animals were killed in inter-group clashes of one sort or another. This is the difficult explaining we have to do, if it is so obvious to the human intellect that it would be better to live peacefully. We describe these killings as men behaving 'like animals', but if we could find a wild animal that showed signs of acting in this way, it would be more precise to describe it as behaving like men. The fact is that we cannot find such a creature. We are dealing with another of the dubious properties that make modern man a unique species.

Biologically speaking, man has the inborn task of defending three things: himself, his family and his tribe. As a pair-forming, territorial, group-living primate he is driven to this, and driven hard. If he or his family or his tribe are threatened with violence, it will be all too natural for him to respond with counter-violence. As long as there is a chance of repelling the attack, it is his biological duty to attempt to do so by any means at his disposal. For many other animals the situation is the same, but under natural conditions the amount of actual physical violence that occurs is limited. It is usually little more than a threat of violence answered by a counter-threat of counter-violence. The more truly violent species all appear to have exterminated themselves—a lesson we should not overlook.

This sounds straightforward enough, but the last few thousand years of human history have over-burdened our evolutionary inheritance. A man is still a man and a family is still a family, but a tribe is no longer a tribe. It is a super-tribe. If we are ever to understand the unique savageries of our national, idealistic and racial conflicts, we must once again examine the nature of this super-tribal condition. We have seen some of the tensions it has set up inside itself—the aggressions of the status battle; now we must look at the way it has created and magnified tensions outside itself, between one group and another.

It is a story of piling on the agony. The first important step was taken when we settled down in permanent dwellings. This gave us a definite object to defend. Our closest relatives, the monkeys and apes, live typically in nomadic bands. Each band keeps to a general home range but constantly moves about inside it. If two groups meet and threaten one another, there is little serious development of the incident. They simply move off and go about their business. Once early man became more strictly territorial, the defence system had to be tightened up. But in the early days there was so much land and so few men that there was plenty of room for all. Even when the tribes grew bigger, the weapons were still crude and primitive. The leaders were themselves much more personally involved in the conflicts. (If only today's leaders were forced to serve in the front lines, how much more cautious and 'humane' they would be when making their initial decisions. It is perhaps not too cynical to suggest that this is why they are still prepared to wage 'minor' wars, but are frightened of major nuclear wars. The range of nuclear weapons has accidentally put them back in the front lines again. Perhaps, instead of nuclear disarmament, what we should be demanding is the destruction of the deep concrete bunkers they have already constructed for their own protection.)

As soon as farming man became urban man, another vital step was taken towards more savage conflict. The division of labour and the specialization that developed meant that one category of the population could be spared for full-time killing—the military was born. With the growth of the urban super-tribes, things began to move more swiftly. Social growth became so rapid that its development in one area easily got out of phase with its progress in another. The more stable balance-of-tribal-power was replaced by the serious instability of super-tribal inequalities. As civilizations flourished and could afford to expand, they frequently found themselves faced, not with equal rivals who would make them think twice and indulge in the ritualized threat of bargaining and trade, but with weaker, more backward groups that could be invaded and assaulted with ease. Flicking through the pages of an historical atlas one can see at a glance the whole sorry story of waste and inefficiency, of construction followed by destruction, only to be followed again by more construction and more destruction. There were incidental advantages, of course, inter-minglings that brought the pooling of knowledge, the spread of new ideas. Ploughshares may have been turned into swords, but the impetus for research into better weapons did lead eventually to better implements as well. The cost, however, was heavy.

As the super-tribes became bigger and bigger, the task of ruling the sprawling, teeming populations became greater, the tensions of overcrowding grew and the frustrations of the super-status race became more intense. There was more and more pent-up aggression, looking for an outlet. Inter-group conflict provided it on a grand scale.

For the modern leader, then, going to war has many advantages that the Stone-Age leader did not enjoy. To start with, he does not have to risk getting his face bloody. Also, the men he sends to their deaths are not personal acquaintances of his: they are specialists, and the rest of society can go about its daily life. Trouble-makers who are spoiling for a fight, because of the super-tribal pressures they have been subjected to, can have their fight without directing it at the super-tribe itself. And having an outside enemy, a villain, can make a leader into a hero, unite his people and make them forget the squabbles that were giving him so many headaches.

It would be naïve to think that leaders are so super-human that these factors do not influence them. Neverthe-less, the major factor remains the urge to maintain or im-prove inter-leader status. The out-of-phase progress of the different super-tribes that I mentioned earlier is undoubted-ly the greatest problem. If, because of its natural resources or its ingenuity, one super-tribe gets one jump ahead of another, then there is bound to be trouble. The advanced group will impose itself on the backward group in one way or another and the backward group will resent it in one way or another. An advanced group is, by its very nature, ex-pansive, and simply cannot bear to leave things alone and mind its own business. It tries to influence other groups, either by dominating them or by 'helping' them. Unless it dominates its rivals to the point where they lose their iden-tities and are absorbed into the advanced super-tribal body (which is often geographically impossible), the situation will be unstable. If the advanced super-tribe helps other groups and makes them stronger, but in its own image, then the day will dawn when they are strong enough to revolt and repel the super-tribe with its own weapons and its own methods.

While all this is going on, the leaders of other powerful, advanced super-tribes will be watching anxiously to make sure that these expansions are not too successful. If they are, then *their* inter-group status will begin to slip.

All this is done under a remarkably transparent but nevertheless persistent cloak of ideology. To read the official documents, one would never guess that it was really the pride and status of the leaders that were at stake. It is always, apparently, a matter of ideals, moral principles, social philosophies or religious beliefs. But to a soldier staring down at his severed legs, or holding his entrails in his hands, it means only one thing: a wasted life. The reason why it was so easy to get him into that position was that he is not only a potentially aggressive animal, but also an intensely co-operative one. All that talk of defending the principles of his super-tribe got through to him because it became a question of helping his friends. Under the stress of war, under the direct and visible threat from the out-group, the bonds between him and his battle companions became immensely strengthened. He killed, more not to let them down than for any other reason. The ancient tribal loyalties were so strong that, when the final moment came, he had no choice.

Given the pressures of the super-tribe, given the global overcrowding of our species, and given the inequalities in progress of the different super-tribes, there is little hope that our children will grow up to wonder what war was all about. The human animal has got too big for its primate boots. Its biological equipment is not strong enough to cope with the unbiological environment it has created. Only an immense effort of intellectual restraint will save the situation now. One sees a sign of this here and there, now and then, but as fast as it grows in one place, it shrivels in another. What is more, we are so resilient as a species that we always seem to be able to absorb the shocks, to make up for the waste, so that we are not even forced to learn from our brutal lessons. The biggest and bloodiest wars we have ever known have done no more, in the long run, than make a tiny, untidy kink in the soaring growth curve of the total world population. There is always a 'post-war bulge' in the birth rate, and the gaps are quickly filled. The human giant regenerates itself like a mutilated flatworm, and slides swiftly on.

What is it that makes a human individual one of 'them', to be destroyed like a verminous pest, rather than one of 'us', to be defended like a dearly beloved brother? What is it that puts him into an out-group and keeps us in the in-group? How do we recognize 'them'? It is easiest, of course, if they belong to an entirely separate super-tribe, with strange customs, a strange appearance and a strange language. Everything about them is so different from 'us' that it is a simple matter to make the gross over-simplification that they are *all* evil villains. The cohesive forces that helped to hold their group together as a clearly defined and efficiently organized society also serve to set them apart from us and to make them frightening by virtue of their unfamiliarity. Like the Shakespearean dragon, they are 'more often feared than seen'.

Such groups are the most obvious targets for the hostility of our group. But supposing we have attacked them and defeated them, what then? Supposing we dare not attack them? Supposing we are, for whatever reason, at peace with other super-tribes for the time being: what happens to our in-group aggression now? We may, if we are very lucky, remain at peace and continue to operate efficiently and constructively within our group. The internal cohesive forces, even without the assistance of an out-group threat, may be sufficiently strong to hold us together. But the pressures and stresses of the super-tribe will still be working on us, and if the internal dominance battle is fought too ruthlessly, with extreme subordinates experiencing too much suppression or poverty, then cracks will soon begin to show. If severe inequalities exist between the sub-groups that inevitably develop within the super-tribe, their normally healthy competition will erupt into violence. Pent-up sub-group aggression, if it cannot combine with the pent-up aggression of other sub-groups to attack a common, foreign enemy, will vent itself in the form of riots, persecutions and rebellions.

Examples of this are scattered throughout history. When the Roman Empire had conquered the world (as it then knew it), its internal peace was shattered by a series of civil wars and disruptions. When Spain ceased to be a conquering power, sending out colonial expeditions, the same thing happened. There is, unhappily an inverse relationship between external wars and internal strife. The implication is clear enough: namely that it is the same kind of frustrated aggressive energy that is finding an outlet in both cases. Only a brilliantly designed super-tribal structure can avoid both at the same time.

It was easy to recognize 'them' when they belonged to an entirely different culture, but how is it done when 'they' belong to our own culture? The language, the customs, the appearance of the internal 'them' is not strange or unfamiliar, so the crude labelling and lumping is more difficult. But it can still be done. One sub-group may not look strange or unfamiliar to another sub-group, but it does look *different*, and that is often enough.

The different classes, the different occupations, the different age-groups, they all have their own characteristic ways of talking, dressing and behaving. Each sub-group develops its own accents or its own slang. The style of clothing also differs strikingly, and when hostilities break out between sub-groups, or are about to break out (a valuable clue), dressing habits become more aggressively and flamboyantly distinctive. In some ways they begin to resemble uniforms. In the event of a full-scale civil war, of course, they actually become uniforms, but even in lesser disputes the appearance of pseudo-military devices, such as arm-bands, badges and even crests and emblems, become a typical feature. In aggressive secret societies they proliferate.

These and other similar devices quickly serve to strengthen the sub-group identity and at the same time make it easier for other groups inside the super-tribe to recognize and lump together the individuals concerned as 'them'. But these are all temporary devices. The badges can be taken off when the trouble is over. The badge-wearers can quickly blend back into the main population. Even the most violent animosities can subside and be forgotten. An entirely different situation exists, however, when a sub-group possesses distinctive *physical* characteristics. If it happens to exhibit, say, dark skin or yellow skin, fuzzy hair or slant eyes, then these are badges that cannot be taken off, no matter how peaceful their owners. If they are in a minority in a super-tribe they are automatically looked upon as a sub-group behaving as an active 'them'. Even if they are a passive 'them' it seems to make no difference. Countless hair-straightening sessions and countless eye-skin-fold operations fail to get the message across, the message that says, 'We are not deliberately, aggressively setting ourselves apart.' There are too many conspicuous physical clues left.

Rationally, the rest of the super-tribe knows perfectly

well that these physical 'badges' have not been put there on purpose, but the response is not a rational one. It is a deep-seated in-group reaction, and when pent-up aggression seeks a target, the physical badge-wearers are there, literally ready-made to take the scapegoat role.

A vicious circle soon develops. If the physical badge-wearers are treated, through no fault of their own, as a hostile sub-group, they will all too soon begin to behave like one. Sociologists have called this a 'self-fulfilling prophecy'. Let me illustrate what happens, using an imaginary example. These are the stages:

1. Look at that green-haired man hitting a child.
2. That green-haired man is vicious.
3. All green-haired men are vicious.
4. Green-haired men will attack anyone.
5. There's another green-haired man—hit him before he hits you.
(The green-haired man, who has done nothing to provoke aggression, hits back to defend himself.)
6. There you are—that proves it: green-haired men *are* vicious.
7. Hit all green-haired men.

This progression of violence sounds ridiculous when expressed in such an elementary manner. It is, of course, ridiculous, but nevertheless it represents a very real way of thinking. Even a dimwit can spot the fallacies in the seven deadly stages of mounting group prejudice that I have listed, but this does not stop them becoming a reality.

After the green-haired men have been hit for no reason for long enough, they do, rather naturally, become vicious. The original false prophecy has fulfilled itself and become a true prophecy.

This is the simple story of how the out-group becomes a hated entity. There are two morals to this tale: do not have green hair; but if you do, make sure you are known personally to people who do not have green hair, so that they will realize that you are not actually vicious. The point is that if the original man seen hitting a child had had no special features potentially setting him apart, he would have been judged as an individual, and there would have been no damaging generalization. Once the harm has been done, however, the only possible hope of preventing a further spread of in-group hostility must be founded on personal interchange and knowledge of the other green-haired individuals *as individuals*. If this does not happen, then the inter-group hostility will harden and the green-haired individuals—even those who are excessively non-violent—will feel the need to club together, even live together, and defend one another. Once this has occurred, then real violence is just around the corner. Less and less contact will take place between members of the two groups and they will soon be acting as if they belonged to two different tribes. The green-haired people will soon start to proclaim that they are proud of their green hair, when in reality it never had the slightest significance for them before it became singled out as a special signal.

The quality of the green-hair signal that made it so potent was its visibility. It had nothing to do with true personality. It was merely an accidental label. No out-group has ever been formed, for example, of people who belong to blood group O, despite the fact that, like skin colour or hair pattern, it is a distinct and genetically controlled factor. The reason is simple enough—you cannot tell who *is* group O, simply by looking at them. So, if a known group O man hits a child, it is difficult to extend antagonism towards him to other group O people.

This sounds so obvious, and yet it is the whole basis of the irrational in-group/out-group hatreds we usually refer to as 'racial intolerance'. For many it is hard to grasp that, in reality, this phenomenon has nothing whatsoever to do with significant racial differences in personality, intelligence, or emotional make-up (which have never been proved to exist), but only with insignificant and nowadays meaningless differences in superficial racial 'badges'. A white child or a yellow child, reared in a black super-tribe and given equal opportunities, would undoubtedly do as well and behave in the same way as the black children. The reverse is also true. If this does not appear to be so, then it is simply the result of the fact that they probably would *not* be given equal opportunities. To understand this we must take a brief look at the way the different races came into being in the first place.

To start with, the word 'race' is unfortunate. It has been abused too often. We speak of the human race, the white race and the British race, meaning respectively the human species, the white sub-species and the British super-tribe. In zoology, a species is a population of animals that breed freely among themselves, but cannot or do not breed with other populations. A species tends to split into a number of distinguishable sub-species as it spreads over a wider and wider geographical range. If these sub-species are artificially mixed up, they still breed freely with one another and can blend back into one overall type, but normally this does not happen. Climatic and other differences influence the colour, shape and size of the different sub-species in their various natural regions. A group living in a cold region, for example, may become heavier and stockier; another, inhabiting a forest region, may evolve a spotted coat that camouflages it in the dappled light. The physical differences help to tune the sub-species in to their environments so that each one is better off in its own particular area. Where the regions meet there is no hard line between the sub-species; they merge gradually into one another. If, as time goes on, they become increasingly different from one another, they may eventually cease to inter-breed at the borders of the range and a sharp dividing line develops. If later they spread and overlap they will no longer mix. They will have become true species.

The human species, as it began to spread out over the globe, started to form distinctive sub-species, just like any other animal. Three of these, the (white) Caucasoid group, the (black) Negroid group and the (yellow) Mongoloid group, have been highly successful. Two of them have not, and exist today as only remnant groups, shadows of their former selves. They are the Australoids—the Australian Aborigines and their relatives—and the Capoids—the southern African bushmen. These two sub-species once covered a much wider range (the bushman at one time owning most of Africa), but they have since been exterminated in all but limited areas. A recent survey of the relative sizes of these five sub-species estimated their present world populations as follows:

Caucasoid: 1,757 million
Mongoloid: 1,171 million
Negroid: 216 million
Australoid: 13 million
Capoid: 126 thousand

Of the total world population of just over 3,000 million human animals, this gives the white sub-species the lead with over 55 per cent, the yellow sub-species close on their heels with 37 per cent and the Negroid sub-species nearly 7 per cent. The two remnant groups together make up less than ½ per cent of the total.

These figures are inevitably approximations, but they give some idea of the general picture. They cannot be precise because, as I explained earlier, the characteristic of a sub-species is that it blends into its neighbours at the places where their ranges meet. An additional complication has arisen in the case of the human species as a result of the increased efficiency of transportation. There has been an enormous amount of migration and shifting about of sub-specific populations, so that in many regions complex mixtures have arisen and a further blending process has taken place. This has occurred despite the formation of in-group/out-group antagonisms and bloodshed because, of course, the different sub-species can still inter-breed fully and efficiently.

Had the different human sub-species remained geographically separated for a longer period of time they might well have split up into distinct species, each physically adapted to its special climatic and environmental conditions. That was the way things were going. But man's increasingly efficient technical control over his physical environment, coupled with his great mobility, has made nonsense of this particular evolutionary trend. Cold climates have been subdued by everything from clothing and log fires to central heating; hot environments have been tamed by refrigeration and air-conditioning. The fact, for instance, that a Negro has more cooling sweat glands than a Caucasoid, is rapidly ceasing to have any adaptive meaning.

In time it is inevitable that the sub-species differences, the 'racial characters', will blend completely and disappear altogether. Our distant successors will stare in wonder at the old photographs of their extraordinary ancestors. Unfortunately this will take a very long time indeed, because of the irrational misuse of these characters as badges for mutual hostility. The only hope of rapidly speeding up this valuable and ultimately inevitable process of re-mixing would be international obedience to a new law forbidding inter-breeding with a member of your own sub-species. Since this is pure fantasy, the solution we must rely on is an increasingly rational approach to what has hitherto been an immensely emotional subject. That this will come easily can soon be refuted by a brief study of the incredible extremes of irrationality that have prevailed on so many occasions. It will suffice to select only one example: the repercussions of the Negro slave trade to America.

Between the sixteenth and the nineteenth centuries a grand total of nearly fifteen million Negroes were captured in Africa and shipped as slaves to the Americas. There was nothing new about slavery, but the scale of the operation and the fact that it was carried out by super-tribes professing the Christian faith made it exceptional. It required a special attitude of mind—one that could only stem from a reaction to the physical differences between the sub-species involved. It could only be done if the African Negroes were looked upon as virtually a new form of domestic animal.

It had not begun like this. The first travellers to penetrate black Africa were astonished by the grandeur and organization of the Negro empire. There were great cities, scholarship and learning, complex administration and considerable wealth. Even today, for many people, this is hard to believe. There is so little evidence of it left, and the propaganda picture of the naked, indolent, murderous savage persists all too well. The glory of the Benin bronzes is easily overlooked. The early reports of Negro civilization have been comfortably hidden away and forgotten.

Let us take just one glimpse at an ancient Negro city in West Africa, as it was seen over three and a half centuries ago by an early Dutch traveller. He wrote:

> The town seemeth to be very great; when you enter into it, you go into a great broad street ... seven or eight times broader than Warmoes street in Amsterdam ... you see many streets on the sides thereof, which also go right forth ... The houses in the town stand in good order one close and even with the other, as the houses in Holland stand ... The King's Court is very great, within it having many great four-square plains, which round about them have galleries ... I was so far within the court that I passed over four such great plains, and wherever I looked, still I saw gates upon gates to go into other places ...

Hardly a crude mud-hut village. Nor could the inhabitants of these ancient West African civilizations be described as ferocious, spear-waving savages. As early as the middle of the fourteenth century a sophisticated visitor remarked on the ease of travel and the reliable availability of food and good lodgings for the night. He commented: 'There is complete security in their country. Neither traveller nor inhabitant in it has anything to fear from robbers or men of violence.'

After the early travellers, the later contacts rapidly turned to commercial exploitation. As the 'savages' were attacked, pillaged, subdued and exported, their civilization crumbled. The remnants of their shattered world began to fit the picture of a barbarous, disorganized race. Reports were more frequent now and they left no doubt as to the inferior nature of the Negroid culture. The fact that this cultural inferiority had been initially caused by white brutality and greed was conveniently overlooked. Instead, the Christian conscience found it easier to accept the idea that the black skin (and the other physical differences) represented outward signs of mental inferiorities. It was then a simple matter to argue that the culture was inferior *because* the Negroes were mentally inferior, and for no other reason. If this was so, then the exploitation did not appear to involve degradation because the 'breed' was already inherently degraded. As the 'proof' rolled in that the Negroes were little better than animals, the Christian conscience could relax.

The Darwinian theory of evolution had not yet arrived on the scene. There were two Christian attitudes towards the existence of Negroid humans: the monogenist and the polygenist. The monogenists believed that all types of men had sprung from the same original source, but that Negroes had long ago undergone a gross physical and moral decline, so that slavery was a proper role for them. Writing in the middle of the last century, an American priest put the position very clearly:

> The Negro is a striking variety, and at present permanent, as the numerous varieties of domestic animals. The Negro will remain what he is, unless his form is altered by intermixture, the simple idea of which is revolting; his intelligence is greatly inferior to that of the Caucasians, and he is consequently, from all we know of him, incapable of governing himself. He has been placed under our protection. The vindication of slavery is contained in the scriptures ... It determines the duties of masters and slaves ... we can effectively defend our institutions from the word of God.

With these words he taunted the early Christian reformers. How dare they go against the Bible?

This statement, coming several centuries after the start of the exploitation, makes it clear just how completely the original knowledge of the ancient civilization of the African

Negroes had been suppressed. If it had not been suppressed, then the 'incapable of governing himself' lie would have been exposed and the whole argument, the whole justification, would have fallen to the ground.

Opposing the monogenists were the polygenists. They believed that each 'race' had been created separately, each with its own special properties, its strengths and its weaknesses. Some polygenists believe that there were as many as fifteen different species of man inhabiting the world. They put in a good word for the Negro:

> The polygenist doctrine assigns to the inferior races of humanity a more honourable place than in the opposite doctrine. To be inferior to another man either in intelligence, vigour, or beauty, is not a humiliating condition. On the contrary, one might be ashamed to have undergone a physical or moral degradation, to have descended the scale of beings, and to have lost rank in creation.

This too was written in the middle of the nineteenth century. Despite the difference in attitude, the polygenists' approach still automatically accepts the idea of racial inferiorities. Either way the Negroes lost out.

Even after the American slaves were given their official freedom, the old attitudes still persisted in one form or another. Had the Negroes not been saddled with their physical out-group 'badges', they would have been rapidly assimilated into their new super-tribe. But their appearance set them apart and the old prejudices were able to persist. The original lie—that their culture had always been inferior and that therefore *they* were inferior—still lurked at the back of white minds. It biased their behaviour and continued to aggravate relations. It influenced even the most intelligent and otherwise enlightened men. It continued to create black resentment, a resentment that was now backed by official social freedom. The outcome was inevitable. Since his inferiority was only a myth, invented by distorting history, the American Negro naturally failed to continue to behave as if he were inferior, once the chains had been removed. He began to rebel. He demanded actual equality as well as official equality.

His efforts were met with staggeringly irrational and violent responses. Real chains were replaced with invisible ones. Segregations, discriminations and social degradations were heaped upon him. This had been anticipated by the early reformers and, at one point in the last century, it was seriously suggested that the whole American Negro population should be 'handsomely rewarded' for their trouble and returned to their native Africa. But repatriation would hardly have returned them to their original civilized condition. That had been smashed long ago. There was no turning back. The damage had been done. They stayed, and tried to gain what was due to them. After repeated frustrations they began to lose their patience, and during the last half-century their revolts have not only persisted but have increased in vigour. Their numbers have risen to around the twenty million mark. They are a force to be reckoned with and Negro extremists have now been driven to a policy, not of simple equality, but of black domination. A second American Civil War seems to be imminent.

Thoughtful white Americans struggle desperately to overcome their prejudice, but the cruel indoctrinations of childhood are difficult to forget. A new kind of prejudice creeps in, an insidious one of over-compensation. Guilt produces an over-friendliness, an over-helpfulness that creates a relationship as false as the one it replaces. It still fails to treat Negroes as individuals. It still persists in looking at them as members of an out-group. The flaw was neatly pin-pointed by an American Negro entertainer who, on being over-enthusiastically applauded by a white audience, chided them by pointing out that they would feel rather foolish if he turned out to be a white man who had blackened his face.

Until human sub-species stop treating other human sub-species as if their physical differences denoted some kind of mental difference, and until they stop reacting to skin colour as if it were being deliberately worn as the badge of a hostile out-group, there will be pointless and wasteful bloodshed. I am not arguing that there can be a world-wide brotherhood of men. That is a naïve utopian dream. Man is a tribal animal and the great super-tribes will always be in competition with another. In well organized societies these struggles will take the form of healthy, stimulating competition and the aggressive rituals of commercial trading and sport, helping to prevent communities from becoming stagnant and repetitive. The natural aggressiveness of men will not become excessive. It will take the acceptable form of self-assertion. Only when the pressures become too great will it boil over into violence.

At either level of aggression—the assertive or the violent—the ordinary (non-racial) in-groups and out-groups will face one another on their own terms. The individuals concerned will not be there by accident. But the situation is entirely different for the individual who, because of the colour of his skin, finds himself accidentally, permanently and inevitably trapped into a particular group. He cannot decide to enter a sub-species group, or leave it. Yet he is treated exactly as though he has become a member of a club, or joined an army. The only hope for the future, as I have said, is that the world-wide mixing up of the originally geographically distinct sub-species, which has been increasingly taking place, will lead to a greater and greater blending of characteristics until the strikingly visible differences have vanished. In the meantime the perpetual need for out-groups on to which in-group aggression can be vented will continue to confuse the issue and will continue to cast alien sub-species in unwarranted roles. Our irrational emotions fail to make the proper distinctions; only the imposition of our rational, logical intellects will help us.

I have selected the example of the American Negro dilemma because it is particularly relevant at the present moment. There is, unhappily, nothing unusual about it. The same pattern has been repeated all over the globe, ever since the human animal became really mobile. Even where there have been no sub-specific differences to fan the flames and keep them alight, extraordinary irrationalities have been widespread. The key error of assuming that a member of another group must possess certain special *inherited* character traits typical of his group, is constantly arising. If he wears a different uniform, speaks a different language or follows a different religion, it is illogically assumed that he also has a *biologically* different personality. Germans are said to be laboriously, obsessively methodical, Italians to be excitedly emotional, Americans to be expansive and extrovert, British to be stiff and retiring, Chinese to be devious and inscrutable, Spaniards to be haughty and proud, Swedes to be bland and mild, French to be querulous and argumentative, and so on.

Even as superficial assessments of *acquired* national characters these generalizations are gross over-simplifications, but they are taken much further: for many people they are accepted as inborn traits of the out-groups concerned. It is really believed that in some way the 'breeds' have come to differ, that there has been some genetical change; but this is nothing more than the illogical wishful thinking of the in-grouping tendency. Confucius put it very well, over two thousand years ago, when he said: 'Men's natures are alike; it is their habits that carry them far apart.' But habits, being mere cultural traditions, can be changed so easily, and the in-grouping urge hopes for something more permanent, more basic, to set 'them' apart from 'us'. Being an ingenious species, if we cannot find such differences, we do not hesitate to invent them. With astonishing aplomb, we airily overlook the fact that nearly all the nations I have mentioned above are complex mixtures of a whole collection of earlier groupings, repeatedly cross-bred and re-fused. But logic has no place here.

The whole human species has a wide range of basic behaviour patterns in common. The fundamental similarities between any one man and any other man are enormous. One of these, paradoxically, is the tendency to form distinct in-groups and to feel that you are somehow different, really deep-down different, from members of other groups. This feeling is so strong that the view I have expressed in this chapter is not a popular one. The biological evidence, however, is overwhelming and the sooner it is appreciated, the more tolerant we can hope to become in our inter-group dealings.

Another of our biological characteristics, as I have already stressed, is our inventiveness. It is inevitable that we shall be constantly trying out new ways of expressing ourselves, and that these new ways will differ from group to group and from epoch to epoch. But these are superficial properties, easily gained and easily lost. They can come and go in a generation, whereas it takes hundreds of thousands of years to evolve a new species like ours and to build its basic biological features. Civilization is only ten thousand years old. We are fundamentally the same animals as our hunting ancestors. We all stemmed from that stock, all of us, regardless of our nationality. We all carry the same basic genetic properties. We are all naked apes beneath the wild variety of our adopted costumes. It is as well for us to remember this when we start playing our in-grouping games and when, under the tremendous pressures of super-tribal living, they begin to get out of hand and we find ourselves about to shed the blood of people who, beneath the surface, are exactly like ourselves.

Having said this, I am nevertheless left with an uneasy feeling. The reason is not hard to find. On the one hand I have pointed out that the in-grouping urge is illogical and irrational; on the other hand I have emphasized that conditions are so ripe for inter-group strife that our only hope is to apply rational, intelligent control. In urging the rational control of the deeply irrational, it could be argued that I am being unduly optimistic. It is not perhaps asking too much that rational processes should be brought to bear as an *aid* to the problem, but on the present evidence it does seem to be beyond hope that they alone will solve it. One only has to obscrvc thc most intellectual of protestors beating policemen over the heads with placards reading 'Stop this violence', or listen to the most brilliant of politicians supporting war 'to ensure peace', to realize that rational restraint in such matters is an elusive quality. Something else is needed. In some way we must tackle, at the roots, those conditions I referred to that are ripening us so effectively for inter-group violence.

I have already discussed these conditions, but it will help to summarize them briefly. They are:

1. The development of fixed human territories.
2. The swelling of tribes into over-crowded super-tribes.
3. The invention of weapons that kill at a distance.
4. The removal of leaders from the front line of battle.

5. The creation of a specialized class of professional killers.
6. The growth of technological inequalities between the groups.
7. The increase of frustrated status aggression within the groups.
8. The demands of the inter-group status rivalrics of the leaders.
9. The loss of social identity within the super-tribes.
10. The exploitation of the co-operative urge to aid friends under attack.

The one condition I have deliberately omitted from this list is the development of differing ideologies. As a zoologist, viewing man as an animal, I find it hard to take such differences seriously in the present context. If one assesses the inter-group situation in terms of actual behaviour, rather than verbalized theorizing, differences in ideology fade into insignificance alongside the more basic conditions. They are merely the excuses, desperately sought for to provide reasons high-sounding enough to justify the destruction of thousands of human lives.

Examining the list of the ten more realistic factors it is difficult to see where one can begin to seek improvement in the situation. Taken together, they appear to offer an absolutely cast-iron guarantee that man will for ever be at war with man.

Remembering that I have described the present state as being that of a human zoo, perhaps there is something we can glean from looking inside the cages of an animal zoo. I have already made the point that wild animals in their natural environment do not habitually slaughter large numbers of their own kind; but what of the caged specimens? Are there massacres in the monkey house, lynchings in the lion house, pitched battles in the bird house? The answer, with obvious qualifications, is in the affirmative. The status struggles between established members of overcrowded groups of zoo animals are bad enough, but, as every zoo-man knows, the situation is even worse when one tries to introduce newcomers to such a group. There is a great danger that the strangers will be jointly set upon and relentlessly persecuted. They are treated as invading members of a hostile out-group. There is little they can do to stem the onslaught. If they huddle unobtrusively in a corner, rather than flaunt themselves in the middle of the cage, they are nevertheless hounded out and attacked.

This does not happen in all instances; where it is most prevalent, the species involved are usually those suffering from the most unnatural degree of spatial cramping. If the established cage owners have more than enough room, they may attack the newcomers initially and drive them away from the favoured spots, but they will not continue to persecute them with undue violence. The strangers are eventually permitted to take up residence in some other part of the enclosure. If the space is too small this stabilization of the relationship can never develop, and bloodshed inevitably ensues.

It is possible to demonstrate this experimetnally. Sticklebacks are small fish that hold territories in the breeding season. The male builds a nest in the water-weeds and defends the area around it against other males of the species. Being solitary in this case, a single male represents the 'in-group' and each of his territory-owning rivals represents an 'out-group'. Under natural conditions, in a river or stream, each male has enough room, so that hostile encounters with rivals are restricted largely to threats and counter-threats. Prolonged battles are rare. If two males are encouraged to build nests, one at either end of a long aquarium tank, then, as in nature, they meet and threaten one another at a roughly mid-tank boundary line. Nothing more violent occurs. However, if the water-weeds in which they have nested have been experimentally planted in small, movable pots, it is possible for the experimenter to shift these pots closer together and artificially cramp the territories. As the pots are gradually brought nearer to one another, the two territory-owners intensify their threat displays. Eventually the system of ritualized threat and counter-threat breaks down, and serious fighting erupts. The males endlessly bite and tear at one another's fins, their nest-building duties forgotten, their world suddenly a riot of violence and savagery. The moment that their nest-pots are drawn apart again, however, peace returns and the battle-ground subsides once more into an arena for harmless, ritualized threat displays.

The lesson is obvious enough: when the small human tribes of early man became swollen into super-tribal proportions, we were, in effect, performing the stickleback experiment on ourselves, with much the same result. If the human zoo is to learn from the animal zoo, then, it is this second condition to which we should pay particular attention.

Viewed with the brutally objective eye of the animal ecologist, the violent behaviour of an over-populated species is an adaptive self-limiting mechanism. It could be described as being cruel to the individual in order to be kind to the species. Each type of animal has its own particular population 'ceiling'. If the numbers rise above this level, some sort of lethal activity intervenes and the numbers sink again. It is worth considering human violence in this light for a moment.

It may sound cold-blooded to express it in this way, but it is almost as if, ever since we first started to become overcrowded as a species, we have been frantically searching for a means to correct this situation and to reduce our numbers to a more suitable biological level. This has not been restricted merely to undertaking bulk slaughter in the form of wars, riots, revolts and rebellions. Our resourcefulness has known no bounds. In the past we have introduced a whole galaxy of self-limiting factors. Primitive societies, when they first began to experience over-crowding, employed practices such as infanticide, human sacrifice, mutilation, head-hunting, cannibalism and all kinds of elaborate sexual taboos. These were not, of course, deliberately planned systems of population control, but they helped to control the population nevertheless. They failed, however, to put a complete brake on the steady increase in human numbers.

As technologies advanced, the individual human life became more strongly protected and these earlier practices were gradually suppressed. At the same time, disease, drought and starvation came under heavy attack. As the populations began to soar, new self-limiting devices appeared on the scene. When the old sexual taboos vanished, strange new sexual philosophies emerged that had the effect of reducing group fecundity; neuroses and psychoses proliferated, interfering with successful breeding; certain sexual practices increased, such as contraception, masturbation, oral and anal intercourse, homosexuality, fetishism and bestiality, which provided sexual consummation without the chance of fertilization. Slavery, imprisonment, castration and voluntary celibacy also played their part.

In addition we terminated individual lives by widespread abortion, murder, the execution of criminals, assassination, suicide, duelling and the deliberate pursuit of dangerous and potentially lethal sports and pastimes.

All these measures have served to eliminate large numbers of human beings from our over-crowded populations, either by the prevention of fertilization, or by extermination. Assembled together in this way they make a formidable list. Yet in the last analysis they have proved, even in combination with mass warfare and rebellion, to be hopelessly ineffectual, The human species has survived them all and has persisted in over-breeding at an ever-increasing rate.

For years there has been a stubborn resistance to interpreting these trends as indications that something is biologically wrong with our population level. We have repeatedly refused to read them as danger signals, warning us that we are heading for a major evolutionary disaster. Everything possible has been done to outlaw these practices and to protect the breeding and living rights of all human individuals. Then, as the groups of human animals have swollen to increasingly unmanageable proportions, we have applied our ingenuity to advancing technologies that help to make these unnatural social conditions bearable.

As each day passes (adding, as it does, so, another 150,000 to the world population), the struggle becomes more difficult. If present attitudes persist, it will soon become impossible. *Something* will eventually arrive to reduce our population level, no matter what we do. Perhaps it will be heightened mental instability leading to reckless utilization of weapons of uncontrollable power. Perhaps it will be mounting chemical pollution, or wildfire diseases of plague intensity. We have a choice: either we can leave matters to chance, or we can attempt to influence the situation. If we take the former course, then there is a very real danger that, when a major population-control factor does break through our defences and starts to operate, it will be like the bursting of a dam and will carry away our whole civilization. If we take the latter course, we may be able to avert this disaster; but how do we set about selecting our method of control?

The idea of enforcing any particular anti-breeding or anti-living device is unacceptable to our fundamentally co-operative nature. The only alternative is to encourage voluntary controls. We could, of course, promote and glamorize increasingly dangerous sports and pastimes. We might popularize suicide ('Why wait for disease?—Die *now*, painlessly!'), or perhaps create a sophisticated new celibacy cult ('purity for kicks'). Advertising agencies throughout the world might be employed to pour out persuasive propaganda extolling the virtues of instant dying.

Even if we took such extraordinary (and biologically wasteful) steps, it is doubtful if they would lead to a significant level of population control. The method more generally favoured today is advanced contraception, with the added secondary measure of legalized abortion in the case of unwanted pregnancies. The argument favouring contraception, as I pointed out in an earlier chapter, is that preventing life is better than curing it. If something has to die, it is better that it should be human eggs and sperm, rather than thinking, feeling human beings, cared for and caring, who have already become an integral and interdependent part of society. If the argument of repugnant waste is applied to contraceived eggs and sperm, it can be pointed out that nature is already remarkably wasteful where these products are concerned, the human female being capable of producing around four hundred eggs during her lifetime, and the adult male literally millions of sperm every day.

There are drawbacks, nevertheless. Just as dangerous sports are likely to eliminate selectively the more adventurous spirits in society, and suicides the more highly strung and imaginative, so contraception may favour a bias against the more intelligent. At their present stage of development, contraceptive devices require a certain level of intelligence, thoughtfulness and self-control if they are to be utilized efficiently. Anyone below that level will be more liable to conceive. If their low level of intelligence is in any way governed by genetic factors, these factors will be passed on to their offspring. Slowly but surely these genetic qualities will spread and increase in the population as a whole.

For modern contraception to work effectively and without bias, therefore, it is essential for urgent progress to be made in the direction of finding less and less demanding techniques; techniques which require the absolute minimum of care and attention. Coupled with this must be a major assault on social attitudes towards contraceptive practices. Only when there are 150,000 fewer fertilizations per day than there are at present, will we be holding the human population steady at its already overgrown level.

Furthermore, although this is difficult enough in itself to achieve, we must add to it the problem of ensuring that the increase in control is suitably spread around the world, rather than concentrated in one or two enlightened regions. If contraceptive advances are unevenly distributed geographically they will inevitably lead to the de-stabilizing of already strained inter-regional relationships.

It is difficult to be optimistic when contemplating these problems, but supposing for the moment they are magically solved and the world population of human animals is holding steady at around its present level of roughly 3,000 million. This means that if we take the whole land surface of the earth and imagine it evenly populated, we are already at a level of more than five hundred times the population density of primitive man. Should we manage to stop the increase and somehow contrive to spread people out more thinly over the globe, we must therefore not delude ourselves that we shall be achieving anything remotely resembling the social condition in which our early ancestors evolved. We shall still require tremendous efforts of self-discipline if we are to prevent violent social explosions and conflicts. But at least we might stand a chance. If, on the other hand, we wantonly allow the population level to go on rising, we shall soon have forfeited that chance.

As if this were not enough, we must also remember that being five hundred times over our natural primitive level is only one of the ten conditions contributing to our present war-like state. It is a frightening prospect, and the danger that we shall completely destroy civilization as we now know it is becoming daily more real.

It is intriguing to contemplate what will happen if we do go. We are making such great strides in the development of ever more efficient techniques of chemical and biological warfare that nuclear weapons may soon become quaintly old-fashioned. Once this has happened, these nuclear devices will gain the respectability of being dubbed conventional weapons and will be tossed recklessly back and forth between the major super-tribes. (With more and more groups joining the nuclear club, the 'hot line' will, of course, by then have become a hopelessly tangled 'hot network'.) The resultant radioactive cloud that will then circle the earth will dispense death to all forms of life in areas that experience rainfall or snowfall. Only the African bushmen and a few other remote groups living in the centres of the most arid desert regions will stand a chance of surviving. Ironically, the bushmen have, to date, been the most dramatically unsuccessful of all human groups and are living still in the primitive hunting condition typical of early man. It looks like being a case of back-to-the-drawing-board, or a supreme example, as someone once predicted, of the meek inheriting the earth.

CHAPTER FIVE

IMPRINTING AND MAL-IMPRINTING

LIVING in a human zoo we have a lot to learn and a lot to remember: but as biological learning machines go, our brain is easily the best in existence. With 14,000 million intricately connected cells churning away, we are capable of assimilating and storing an enormous number of impressions.

In everyday use the machinery runs very smoothly, but when something exceptional occurs in the outside world we switch on to a special emergency system. It is then that, in our super-tribal condition, things can go astray. There are two reasons for this. On the one hand, the human zoo in which we live shields us from certain experiences. We do not regularly kill prey—we buy meat. We do not see dead bodies—they are covered by a blanket or hidden in a box. This means that when violence does break through the protective barriers, its impact on our brains is greater than usual. On the other hand, the kinds of super-tribal violence that do break through are frequently of such unnatural magnitude that they are painfully impressive, and our brains are not always equipped to deal with them. It is this type of emergency learning that deserves more than a passing glance here.

Anyone who has ever been involved in a serious road accident will understand what I mean. Every tiny, nasty detail is, in a flash, burned into the memory and stays there for life. We all have personal experiences of this sort. At the age of seven, for example, I was nearly drowned, and to this day I can recall the incident as vividly as if it were yesterday. As a result of this childhood experience, it was thirty years before I was able to force myself to conquer my irrational fears of swimming. Like all children I had many other unpleasant experiences during the course of growing up, but the vast majority of them left no lasting scars.

It seems, then, that as we go through our lives we encounter two different kinds of experience. In one type, brief exposure to a situation makes an indelible and unforgettable impact; in the other, it produces only a mild and easily forgotten impression. Using the terms rather loosely, we can refer to the first as involving traumatic learning, the second as involving normal learning. In traumatic learning the effect produced is out of all proportion to the experience that caused it. In normal learning, the original experience has to be repeated over and over again to keep its influence going. Lack of reinforcement of ordinary learning leads to a fading of the response. In traumatic learning it does not.

Attempts to modify traumatic learning meet with enormous difficulties and can easily make matters worse. In normal learning this is not so. My drowning incident illustrates the point. The more I was shown the pleasures of swimming, the more intense my hatred of it became. If the early incident had not had such a traumatic effect I would have responded more and more positively instead of more and more negatively.

Traumas are not the main subject of this chapter, but they make a useful introduction to it. They show clearly that the human animal is capable of a rather special kind of learning, a kind that is incredibly rapid, difficult to modify, extremely long-lasting and requires no practice to keep it perfect. It is tempting to wish that we could read books in this way, remembering their entire contents for ever after only a single, brief scanning. However, if all our learning worked like this, we would lose all sense of values. Everything would have equal importance and we would suffer from a serious lack of selectivity. Rapid, indelible learning is reserved for the more vital moments of our lives. Traumatic experiences are only one side of this coin. I want to turn it over and examine the other side, the side that has been labelled 'imprinting'.

Whereas traumas are concerned with painful, negative experiences, imprinting is a positive process. When an animal experiences imprinting it develops a positive attachment to something. As with traumatic experiences, the process is quickly over, almost irreversible and needs no later reinforcement. In human beings it happens between a mother and her child. It can happen again when the child grows up and falls in love. Becoming attached to a mother, a child, or a mate are three of the most vital bits of learning that we can undergo in our entire lives and it is these that have been singled out for the special assistance that the phenomenon of imprinting gives. The word 'love' is, in fact, the way we commonly describe the emotional feelings that accompany the imprinting process. But before we go deeper into the human situation, a brief look at some other species will be helpful.

Many young birds, when they hatch from the egg, must immediately form an attachment to their mother and learn to recognize her. They can then follow her around and keep close to her for safety. If newly hatched chicks or ducklings did not do this, they might easily become lost and perish. They are too active and mobile for the mother to be able to keep them together and protect them without the assistance of imprinting. The process can take place in literally a matter of minutes. The first large moving object that the chicks or ducklings see, automatically becomes 'mother'. Under normal conditions, of course, it really is their mother, but in experimental situations it can be almost anything. If the first large moving object that incubator chicks see happens to be an orange balloon, pulled along on a piece of string, then they will follow that. The balloon quickly becomes 'mother'. So powerful is this imprinting process, that if, after some days, the chicks are then given a choice between their adopted orange balloon and their real mother (who has previously been kept out of sight), they will prefer the balloon. No more striking proof of the imprinting phenomenon can be provided than the sight of a batch of experimental chicks eagerly pattering along in the wake of an orange balloon and completely ignoring their genuine mother near by.

Without experiments of this kind, it could be argued that the young birds become attached to their natural mother because they are rewarded by being with her. Staying close to her means keeping warm, finding food, water and so on. But orange balloons lead to no such rewards and yet they easily become powerful mother-figures. Imprinting, then, is not a matter of responding to rewards, as in ordinary learning. It is simply a matter of exposure. We could call it 'exposure learning'. Also, unlike most normal learning, it has a critical period. Young chicks and ducklings are sensitive to imprinting for only a very brief period of days after hatching. As time passes they become frightened of large moving objects and, if not already imprinted, find it difficult to become so.

As they grow up, the young birds become independent and cease to follow the mother. But the impact of the early imprinting has not been lost. It has not only told them who their mother was, it has also told them what species they belong to. As adults, it helps them to select a sexual partner from their own species rather than from some alien species.

Again, this has to be proved by experiments. If young

animals of one species are reared by foster parents of another species, then, when they mature, they may try to mate with members of the foster species instead of with their own kind. This does not always occur, but there are many examples of it. (We still do not know why it occurs in some cases but not in others.)

Among captive animals this susceptibility to becoming fixated on the wrong species can lead to some bizarre situations. When doves reared by pigeons become sexually mature, they ignore other doves and try to mate with pigeons. Pigeons reared by doves try to mate with doves. A zoo peacock, reared on its own in a giant tortoise enclosure, displayed persistently to the bewildered reptiles, refusing to have anything to do with newly arrived peahens.

I have called this phenomenon 'mal-imprinting'. It occurs widely in the world of man/animal relationships. When certain animals, isolated from their own kind from birth, are hand-reared by human beings, they may later respond, not by biting the hand that fed them, but by copulating with it. Doves have often been found to react like this. It is not a new discovery. It has been known from ancient times, when Roman ladies kept small birds to amuse themselves in this way. (Leda, it seems, was more ambitious.) Pet mammals sometimes clasp and attempt to copulate with human legs, as certain dog-owners are painfully aware. Zoo keepers also have to keep a wary eye open during the breeding season. They must be ready to resist the advances of everything from an amorous emu to a rutting deer, when members of these species have been isolated and hand-reared. I myself was once the embarrassed recipient of sexual advances by a female giant panda. It occurred in Moscow, where I had arranged for her to be taken to be mated with the only male giant panda outside China. She ignored his persistent sexual attentions, but when I put my hand through the bars and patted her on the back, she responded by raising her tail and directing a full sexual invitation posture at me, while the male panda was only a few feet away. The difference between the two animals was that she had been isolated from other pandas at a much earlier age than the male. He had matured as a panda's panda, but she was now a people's panda.

Sometimes a 'humanized' animal may appear to be able to tell the difference between a human male and a human female, when making sexual advances to them, but this can be deceptive. A mal-imprinted male turkey, for instance, tried to mate with men, but attacked women. The reason was an intriguing one. Women wear skirts and carry handbags. Aggressive male turkeys display with drooping wings and with wattles. In the eyes of the mal-imprinted male turkey, the skirts became drooping wings and the handbags became wattles. It therefore saw women as rival males and attacked them, reserving its sexual advances for men.

Zoos are full of animals that, with misguided human kindness, have been painstakingly nurtured and hand-reared and then returned to the company of their own kind. But as far as tame isolates are concerned, their own kind are now aliens, members of some frightening, strange, 'other' species. There is an adult male chimpanzee at one zoo that has been caged with a female for over ten years. Medical tests show that he is sexually healthy and she is known to have bred before she was put with him. But because he was a hand-reared isolate, he ignores her completely. He never sits with her, grooms her or attempts to mount her. As far as he is concerned, she belongs to another species. Years of exposure to her have not changed him.

Such animals may become extremely aggressive towards their own species, not because they are treating them as rivals, but because they see them as foreign enemies. The usual rituals, that under normal circumstances make for bloodless settlements, break down. A female mongoose, hand-reared and tame, was given a wild-caught male in the hope that they would breed, but she attacked him from the moment he entered the cage. Eventually they appeared to reach a state of fairly stable mutual disagreement, but the male must have been under considerable stress because he soon developed ulcers and died. The female immediately became her old friendly self again.

A hand-reared tigress was placed in a cage next to a wild-caught male for the first time in her life. She could see him and smell him, but they could not yet meet. This was just as well. She was so 'humanized' that as soon as she detected his presence, she fled to the far side of her cage and refused to move. This was an abnormal reaction for a tigress, but a much more normal one for a member of her adopted (human) species on encountering a tiger. She went further: she stopped eating, and continued to refuse food for several days, until the male was taken away. It took several weeks in her case to bring her back to her normally friendly, active self again, rubbing up against the bars to be patted and fondled by her keepers.

Sometimes the rearing conditions are such that the animal develops a dual sexual personality. If it is reared by humans in the presence of other members of its own species it may, as an adult, try to mate both with humans and with its own kind. The mal-imprinting is only partial, there being some degree of normal imprinting as well. This would be unlikely with a very rapid imprinter such as a duckling or a chick, but mammals tend to become socialized more slowly. There is time for a dual imprint to occur. Careful American studies with dogs have shown this very clearly. The socialization phase for domestic dogs lasts from the age of twenty days to sixty days. If domestic puppies are completely isolated from man (being fed by remote control) throughout this period, they emerge at the other end as virtually wild animals. If, however, they are reared in the presence of both dogs and men, then they are friendly towards both.

Monkeys reared in total isolation, both from other monkeys and other species including man, find it almost impossible to adapt in later life to any kind of social contact. Placed with sexually active members of their own species, they do not know how to respond. Most of the time they are terrified of making any social contact and sit nervously in a corner. They are so un-imprinted that they are virtually non-social animals, even though they belong to a highly sociable species. If they are reared with other young animals of their own kind, but without mothers, they do not suffer in this way, so that there also appears to be a kind of companion-imprinting as well as a parental one. Both processes can play a part in attaching an animal to its species.

The world of the mal-imprinted animal is a strange and frightening place. Mal-imprinting creates a psychological hybrid, performing patterns of behaviour belonging to its own species, but directing them to its adopted foster species. Only with enormous difficulty, and sometimes not even then, can it re-adapt. For some species the sexual signals of its own kind are strong enough, the responses to them instinctive enough, to enable it to survive its abnormal upbringing, but for many the power of imprinting is such that it overrides everything.

Animal lovers would do well to remember this when indulging in the 'taming' of young wild animals. Zoo officials have long been perplexed by the great difficulties they have encountered in breeding many of their animals. Sometimes this has been due to inadequate housing or feeding, but all too frequently the cause has been mal-imprinting before the animals concerned arrived at the zoo.

Turning to the human animal, the significance of imprinting is clear enough. During the early months of its life a human baby passes through a sensitive socialization phase when it develops a profound and long-lasting attachment to its species and especially to its mother. As with animal imprinting, the attachment is not totally dependent on physical rewards obtained from the mother, such as feeding and cleaning. The exposure learning typical of imprinting also takes place. The young baby cannot keep close to the mother by following her like a duckling, but it can achieve the same end by the use of the smiling pattern. Smiling is attractive to the mother and encourages her to stay with the infant and play with it. These playful, smiling interludes help to cement the bond between the child and its mother. Each becomes imprinted on the other and a powerful reciprocal attachment develops, a persistent bond that is extremely important for the later life of the child. Infants that are well fed and cleaned, but are deprived of the 'loving' of early imprinting, can suffer anxieties that stay with them for the rest of their lives. Orphans and babies which have to live in institutions, where personal attention and bonding are unavoidably limited, all too frequently become anxious adults. A strong bond cemented during the first year of life will mean a capacity for making strong bonds during the adult life that follows.

Good early imprinting opens a large emotional bank account for the child. If expenses are heavy later on, it will have plenty to draw upon. If things go wrong with its parental care as it grows up (such as parental separation, divorce or death), its resilience will depend on the attachment quality of that first vital year. Later troubles will, of course, take their toll, but they will be minor compared with troubles in the early months. A five-year-old child, evacuated from London during the last war and separated from its parents, when asked who he was, replied: 'I'm nobody's nothing.' The shock clearly was damaging, but whether in such cases it will cause lasting harm will depend to a large extent on whether it is confirming or contradicting earlier experiences. Contradiction will cause bewilderment that can be rectified, but confirmation will tend to harden and strengthen earlier anxieties.

Passing on to the next great attachment phase, we come to the sexual phenomenon of pair-bonding. 'Falling in love at first sight' may not happen to all of us, but it is far from being a myth. The act of falling in love has all the properties of an imprinting process. There is a sensitive period (early adult life) when it is most likely to occur; it is a relatively rapid process; its effect is long-lasting in relation to the time it takes to devlop; and it is capable of persisting even in the conspicuous absence of rewards.

Against this it might be argued that for many of us the earliest pair-bondings are unstable and ephemeral. The answer is that during the years of puberty and immediate post-puberty, the capacity to form a serious pair-bond takes some time to mature. This slow maturation provides a transition phase during which we can, so to speak, test the water before jumping in. If it were not so, we would all become completely fixated on our first loves. In modern society the natural transition phase has been artificially lengthened by the undue persistence of the parental bond. Parents tend to cling on to their offspring at a time when, biologically speaking, they should be releasing them. The reason is straightforward enough: the complex demands of the human zoo make it impossible for a fourteen- or fifteen-year-old individual to survive independently. This inability imparts a child-like quality which encourages the mother and father to continue to respond parentally despite the fact that their offspring is now sexually mature. This in turn prolongs many of the offspring's infantile patterns, so that they overlap unnaturally with the new adult patterns. Considerable tensions arise as a result and there is often a clash between the parent/offspring bond and the freshly developing tendency in the young to form a new sexual pair-bond.

It is not the parents' fault that their children cannot yet fend for themselves in the super-tribal world outside; nor is it the children's fault that they cannot avoid transmitting infantile signals of helplessness to their parents. It is the fault of the unnatural urban environment, which requires more years of apprenticeship than the biological growth rate of the young human animal provides.

Despite this interference with the development of the new pair-bond relationship, sexual imprinting soon forces its way to the surface. Young love may be typically ephemeral, but it can also be extremely intense—so much so that permanent fixations on 'childhood sweethearts' do occur in a number of cases, regardless of the socio-economic impracticability of the relationships. Even if, under pressure, these early pair-bonds collapse, they can leave their mark. Frequently it seems as if the later search for a sexual partner, at the fully independent adult phase, involves an unconscious quest to re-discover some of the key characteristics of the very first sexual imprint. Ultimate failure in the quest may well be a hidden factor helping to undermine an otherwise successful marriage.

This phenomenon of *bond confusion* is not confined to the 'childhood sweetheart' situation. It can occur at any stage, and is particularly likely to harass second marriages, where silent and sometimes not-so-silent comparisons with earlier mates are frequently being made. It can also play another important and damaging role when the parent/offspring bond is confused with the sexual pair-bond. To understand this it is necessary to look again at what the parent/offspring bond does to the infant. It tells it three things: 1. This is my particular, personal parent. 2. This is the species I belong to. 3. This is the species I shall mate with in later life.

The first two instructions are straightforward; it is the third that can go wrong. If the early bond with the parent of the opposite sex has been particularly persistent, some of their *individual* characteristics can also be carried over to influence the later sexual bonding of the offspring. Instead of taking in the message as 'This is the species I shall mate with in later life,' the child reads it as 'This is the type of person I shall mate with in later life.'

A limiting influence of this kind can become a serious problem. Interference with the sexual pair-bonding process, stemming from a persistent parental image, can lead to a particular mate-selection which, in all other respects, is highly unsuitable. Conversely, an otherwise thoroughly compatible mate can fail to achieve a full relationship because he or she lacks certain trivial but key characteristics of the partner's parent. ('My father would never do that.'—'But I'm not your father.')

This troublesome phenomenon of bond confusion appears to be caused by the unnatural levels of family-unit isolation that so often develop in the crowded world of the human zoo. The 'strangers-in-our midst' phenomenon tends to clamp down on the tribal-sharing, social mixing atmosphere typical of smaller communities. Defensively, the families turn in on themselves, boxing themselves off from one another in neat rows of terraced or semi-detached cages. Unhappily there is no sign of the situation improving: rather the reverse.

Leaving the question of bond confusion, we now have to consider another, stranger aberration of human imprinting: our own version of mal-imprinting. Here we enter the unusual world of what has been termed sexual fetishism.

For a minority of individuals the nature of the first sexual experience can have a psychologically crippling effect. Instead of becoming imprinted with the image of a particular mate, this type of individual becomes sexually fixated on some inanimate object present at the time. It is not at all clear why so many of us escape these reproductively abnormal fixations. Perhaps it depends on the vividness or violence of certain aspects of the occasion of our first major sexual discovery. Whatever it is, the phenomenon is a striking one.

Judging by the case-history records that are available, it appears that the attachment to a sexual fetish occurs most frequently when the initial sexual consummation takes place spontaneously or when the individual is alone. In many instances it can be traced to the first ejaculation of a young adult male, which often occurs in the absence of a female and without the usual pair-bonding preliminaries. Some characteristic object that is present at the moment of ejaculation instantly takes on a powerful and lasting sexual significance. It is as if the whole imprinting force of pair-bonding is accidentally channelled into an inanimate object, giving it, in a flash, a major role for the rest of the person's sexual life.

This striking form of mal-imprinting is probably not quite so rare as it seems. Most of us develop a primary pair-bond with a member of the opposite sex, rather than with fur gloves or leather boots, and we are happy to advertise our pair-bonds openly, confident that others will understand and share our feelings; but the fetishist, firmly imprinted with his unusual sexual object, tends to remain silent on the subject of his strange attachment. The inanimate object of his sexual imprinting, which has such enormous significance for him, would mean nothing to others and, for fear of ridicule, he keeps it secret. It not only means nothing to the vast majority of people, the non-fetishists, but it also means little to other fetishists, each having their own particular speciality. Fur gloves have as little significance for a leather-boot fetishist as they do for a non-fetishist. The fetishist therefore becomes isolated by his own, highly specialized form of sexual imprinting.

Against this, it can be said that there are certain kinds of objects that do crop up with striking frequency in the fetish world. Rubber goods are particularly common, for instance. The significance of this will become clearer if we examine a few specific cases of fetish development.

A twelve-year-old boy was playing with a fox-fur coat when he experienced his first ejaculation. In adult life he was only able to achieve sexual satisfaction in the presence of furs. He was unable to copulate with females in the ordinary way. A young girl experienced her first orgasm when clutching a piece of black velvet as she masturbated. As an adult, velvet became essential to her sexually. Her whole house was decorated with it and she only married in order to obtain more money to buy more velvet. A fourteen-year-old boy had his first sexual experience with a girl who was wearing a silk dress. Later, he was incapable of making love to a naked female. He could only become aroused if she was wearing a silk dress. Another young boy was leaning out of a window when his first ejaculation occurred. As it happened, he saw a figure moving past in the road outside, walking on crutches. When he was married he could only make love to his wife if she wore crutches in bed. A nine-year-old boy was playing with a soft glove against his penis at the moment of his first ejaculation. As an adult he became a glove-fetishist with a collection of several hundred gloves. All his sexual activities were directed to these gloves.

There are many examples of this kind, clearly linking the adult fetish to the first sexual experience. Other common fetish objects include: shoes, riding boots, stiff collars, corsets, stockings, underclothes, leather, rubber, aprons, handkerchiefs, hair, feet and special costumes such as nurse-maids' uniforms. Sometimes these become the essential elements necessary for a successful (and otherwise normal) copulation. Sometimes they completely replace the sexual partner. Texture appears to be an important feature of most of them, often because pressures and frictions of various kinds are significant in causing the first sexual arousal in an individual's lifetime. If some substance with a highly characteristic tactile quality is involved, then it seems to stand a strong chance of becoming a sexual fetish. This could account for the high frequency of rubber, leather and silk fetishes, for example.

Shoe, boot and foot fetishes are also common and it seems likely that here, too, a pressure against the body could easily be involved. There is one classic case of a fourteen-year-old boy who was playing with a twenty-year-old girl who was wearing high-heeled shoes. He was lying on the ground and she playfully stood on him and trampled him. When her foot rested on his penis he experienced his first ejaculation. As an adult this became his only form of sexual activity. During his life he managed to persuade over a hundred women to trample on him wearing high-heeled shoes. Ideally the partner had to be of a particular weight and the shoes of a particular colour. The original encounter had to be re-created as precisely as possible to produce a maximum reaction.

This last case shows very clearly how masochism can develop. Another young boy, for example, had his first sexual experience spontaneously while wrestling with a much larger girl. In later life he was fixated on heavy, aggressive women who were prepared to hurt him during sexual encounters. It is not difficult to imagine how certain forms of sadism could develop in a similar way.

The attachment to a sexual fetish differs from the process or ordinary conditioning in several ways. Like imprinting (or the traumatic experiences I mentioned at the beginning of the chapter) it is very rapid, has a lasting effect and is extremely difficult to reverse. It also appears at a sensitive period. Like mal-imprinting, it fixes the individual on to an abnormal object, channelling sexual behaviour away from the biologically normal object, namely a member of the opposite sex. It is not so much the positive acquiring of sexual significance of an object such as a rubber glove that causes the damage; it is the total elimination of all other sexual objects that creates the problem. The mal-imprinting is so powerful in the cases I have mentioned that it 'uses up' all the available sexual interest. Just as the experimental duckling will follow only the orange balloon and completely ignore its real mother, so the glove fetishist will only mate with a glove and completely ignores potential mates. It is the exclusivity of the imprinting process that causes the difficulties when the mechanism misfires. We all find various textures and pressures stimulating as accessories to sexual encounters. There is nothing strange about responding to soft silks and velvets. But if we become exclusively fixated on them, so that we develop what amounts to a pair-bond with them (like the shoe-fetishist who, when he was alone with girls' shoes, 'blushed in their presence as if they were the girls themselves'), then something has gone drastically wrong with the imprinting mechanism.

Why should a small, but nevertheless considerable number of human animals suffer from this kind of mal-imprinting? Other animals, under natural conditions in the wild, do not appear to do so. For them it only occurs when they are caught and hand-reared under highly artificial conditions, or when they are kept in enclosures with alien species, or when special experiments are carried out. Perhaps this gives the clue. As I have already emphasized, in a human zoo social conditions are highly artificial for our simple tribal species. In many of our super-tribes sexual behaviour is severely restricted at the critical stage of puberty. But although it becomes hidden and cloaked with all kinds of unnatural inhibitions, nothing can hold it back completely. It soon bursts through. If, when it does so, there are certain highly characteristic objects present, then they may become over-impressive. Had the developing adolescent become gradually more experienced in sexual matters at an earlier stage and had his initial sexual explorations been richer and less constricted by the artificialities of the super-tribe, then the later mal-imprinting could perhaps have been avoided. It would be interesting to know how many of the extreme fetishists were solitary children without brothers and sisters, or, as young adolescents, were timid and shy of making personal contacts, or lived in a rather strict household. Future research is needed here but my guess is that the proportion would be high.

One important form of mal-imprinting that I have not mentioned is homosexuality. I have left it until now because it is a more complex phenomenon and because mal-imprinting is only part of the story. Homosexual behaviour can arise in one of four ways. Firstly, it can occur as a case of mal-imprinting in much the same way as fetishism. If the earliest sexual experience in an individual's life is a powerful one and occurs as a result of an intimate encounter with a member of the same sex, then a fixation on that sex can rapidly develop. If two adolescent boys are wrestling together or indulging in some form of sex play, and ejaculation occurs, this can lead to mal-imprinting. The strange thing is that boys often share early sexual experiences of one kind or another and yet the majority survive to become adult heterosexuals. Again we need to know much more about what it is that fixates a few but not the majority. As with the fetishists, it probably has something to do with the degree of the richness of the boy's social experience. The more restricted he has been socially, the more cut off from personal interactions, the blanker will be his sexual canvas. Most boys have, as it were, a sexual blackboard on which things are lightly sketched, rubbed out and re-drawn. But the inward-living boy keeps his sexual canvas virginally white. When finally something does get drawn on it, it will have a much more dramatic impact and he will probably keep the picture for life. Rough-and-tumble, extrovert boys may become involved in homosexual activities, but they will simply put them down to experience and pass on, adding more and more experiences as they progress with their socializing explorations.

This leads me to the other causes of persistent homosexual behaviour. I say 'persistent' because, of course, brief and fleeting homosexual activities occur for the vast majority of both sexes at some point in their lives, as part of general sexual explorations. For most people, like the rough-and-tumble boys, they are mild experiences and are usually confined to childhood. But for some, homosexual patterns persist throughout life, frequently to the almost total, or total, exclusion of heterosexual activities. Mal-imprinting of the type I have been discussing does not explain all these cases. A second, very simple cause is that the *opposite* sex behaves in an exceptionally unpleasant way towards a particular individual. A boy terrorized by girls may well come to regard other males as more attractive sexual partners, despite the fact that, as mates, they are sexually inadequate objects. A girl terrorized excessively by boys may react in the same way and turn to other girls as sexual partners. Terrorizing is not the only mechanism, of course: betrayal, and other forms of social or physical punishment from the opposite sex, can work just as effectively. (Even if the opposite sex is not directly hostile, cultural pressures placing powerful restrictions on heterosexual activities may lead to the same result.)

A third major influence in the creation of a persistent homosexual is the childhood assessment of the roles of his or her parents. If a child has a weak father who is dominated by the mother, it is particularly likely to get the masculine and feminine roles confused and reversed. This then tends to lead to a choice of the wrong sex as a pair-bond partner in later life.

The fourth cause is a more obvious one. If members of the opposite sex are totally absent from the environment for a long period of time, then members of the same sex become the next best thing for sexual encounters. A male isolated from females in this way, or a female isolated from males, may persistently indulge in homosexuality without any of the other three factors I have mentioned having any influence at all. A male prisoner, for instance, may have escaped mal-imprinting, may be fond of the opposite sex and may have had a father who dominated his mother in a completely masculine way, and yet he may still become a long-term homosexual if he is confined in an all-male prison community, where the nearest thing to a female body is another male body. If, in prisons, in boarding-schools, on naval vessels or in army barracks, the uni-sexual condition lasts for some years, the opportunist homosexual may eventually become conditioned to the rewards of his enforced sexual patterns and may persist in them even after he has returned to a heterosexual environment.

Of these four influences leading to persistent homosexual behaviour, only the first one is relevant to the present chapter, but it was important to discuss them all here in order to explain the partial role that mal-imprinting plays in this particular sexual phenomenon.

Homosexual behaviour in other animals is usually of the next-best-thing variety, and disappears in the presence of sexually active members of the opposite sex. There are a few cases of persistently homosexual animals, however, in instances where special social experiments have been carried out. If young mallard ducklings, for example, are kept in all-male groups of from five to ten individuals for the first seventy-five days of their lives, and never encounter a female of their species during that time, they become permanently homosexual. When released on to a pond as adults, now with both males and females present, they completely ignore the females and set up homosexual pair-bonds between themselves. This situation persists for many years, probably the whole life-time of the homosexual ducks, and nothing the females can do will alter it. Doves kept in homosexual pairs are well known to copulate with one another and may form complete pair-bonds. Two males that became sexually imprinted on one another in this way went through the whole breeding cycle together, co-operating to build a nest, incubate eggs and rear the young. The fertile eggs, of course, had to be provided from the nest of a true pair, but they were quickly accepted, each of the homosexual males reacting as if they had been laid by his partner. If a real female had been introduced after the homosexual pair-bond had launched the two males into their pseudo-reproductive cycle, it is doubtful if they would have taken any notice of her. By that stage the homosexuality would have become persistent, at least for the duration of that complete breeding cycle.

Mal-imprinting in the human animal is not confined to sexual relationships. It can also occur in the parent-offspring relationships. As far as human infants becoming imprinted on parents of the wrong species is concerned, good evidence is lacking. The famous cases of so-called 'wolf-children' (abandoned or lost babies being suckled and reared by wolf bitches) have never been fully substantiated and must remain for the time being in the realm of fiction. If such a thing could occur, however, there is little doubt that the wolf-children would become fully mal-imprinted on their foster-parents.

The reverse process, by contrast, is encountered almost every day. When a young animal is hand-reared by a human foster-parent, it is not only the pet animal that becomes mal-imprinted. The human foster-parent also often becomes intensely mal-imprinted and responds to the young animal as if it were a human baby. The same kind of emotional devotion is lavished on it and the same kind of heartbreaks occur if something goes wrong.

Just as a pseudo-parent, such as the duckling's orange balloon, has certain key qualities that make it suitable for mal-imprinting (it is a large moving object), so the pseudo-infant becomes more suitable if it possesses certain qualities typical of the human infant. Human babies are helpless, soft, warm, rounded, flat-faced, big-eyed and they cry. The more of these properties a young animal possesses, the more likely it is to encourage the setting up of a parent-offspring bond with a mal-imprinted human foster parent. Many young mammals have nearly all these properties and it is extremely easy for a human being to become mal-imprinted with them in a matter of minutes. A soft, warm, big-eyed fawn bleating for its mother, or a helpless, rounded puppy crying for a missing bitch, projects a powerful infantile image which few human females can resist. Since some of the childlike properties of such animals are even stronger than those of a real human baby, the exaggerated stimuli from the pseudo-infant can frequently become more powerful than the natural ones, and the mal-imprinting becomes intense.

Animal pseudo-infants have one big drawback: they grow up too quickly. Even slow developers become active adults in only a fraction of the time it takes for a real human infant to mature. When this happens they often become unmanageable and lose their appeal. But the human animal is an ingenious species and has taken steps to deal with this unfortunate development. By selective breeding over a period of centuries, it has managed to make its domestic pets more infantile, so that adult cats and dogs, for example, are rather juvenile versions of their wild counterparts. They remain more playful and less independent, and continue to fulfil their roles as child substitutes.

With some breeds of dogs (the lap dogs or 'toy' dogs) this process has been taken to extremes. They not only behave in a more juvenile way, they also look, feel and sound more juvenile. Their whole anatomy has been altered to make them fit more closely to the image of a human baby, even when they are adult. In this way they can act as a satisfying pseudo-infant, not just for a few months as puppies, but for ten years or longer, a time span that begins to match that of human childhood. What is more, they go one better than the real baby, because they remain baby-like throughout the whole period.

The Pekinese is a good example. The wild ancestor of the Pekinese (as of all domestic dogs) is the wolf, a creature that can weigh up to 150 pounds or more. The average weight of an adult European human is much the same, about 155 pounds. The weight of a newborn human baby is roughly between five and ten pounds, the average being slightly over seven pounds. So, to convert the wolf into a good pseudo-infant, it has to be reduced in size to about one-fifteenth of its original, natural weight. The Pekinese is a triumph of this process, weighing today between seven and twelve pounds, with an average of about ten pounds. So far, so good. It matches the baby in weight and, even as an adult, has the first of the vital pseudo-infant properties: it is a small object. But some other improvements are needed. The legs of a typical dog are too long in relation to its body. Their proportion is more reminiscent of the human adult than the short-limbed human baby. So, off with their legs! By careful selective breeding it is possible to produce strains with shorter and shorter legs until they can only waddle along. This not only corrects the proportions, but as a bonus it also renders the animals more clumsy and helpless. Again, valuable infantile features. But something is still missing. The dog is warm enough to the touch, but not soft enough. Its natural wild-type hair is too short, stiff and coarse. So, on with the hair! Selective breeding again comes to the rescue, producing long, soft, flowing silky hair, creating the essential feel of infantile super-softness.

Further modifications are necessary to the natural wild shape of the dog. It has to become plumper, bigger-eyed and shorter-tailed. One only has to look at a Pekinese to see that these changes have also been successfully imposed. Its ears stuck up and were too pointed. By the device of making them bigger, floppier and covered in long flowing hair, it was possible to convert them into a reasonable semblance of a growing infant's hair-style. The voice of the wild wolf is too deep, but the reduction in body size has taken care of that, producing a higher-pitched, more infantile tone. Finally, there is the face. A wild dog's face is far too pointed, and a little genetic plastic surgery is needed here, too. No matter if it deforms the jaws and makes feeding difficult, it has to be done. And so the Pekinese has its face squashed flat and childlike. Again there is an added bonus, because this also makes it more helpless and more dependent on its pseudo-parent for providing suitably prepared food, another essential parental activity. And there sits our Pekinese pseudo-infant, softer, rounder, more helpless, bigger-eyed and flatter-faced, ready to set up a powerful mal-imprinted bond in any susceptible adult human who happens along. And it works. It works so well that they are not only mothered, but also live with humans, travel with them, have their own (veterinary) doctors and are frequently buried in graves like humans and even left money in wills like real human offspring.

As I have said before, on other topics, this is a description, not a criticism. It is difficult to see why so many people criticize such activities when they so obviously fulfil a basic need that often cannot be satisfied in the normal way. It is even harder to see why some people can accept this kind of imprinting, but not other kinds. Many humans are repelled by sexual mal-imprinting, for example, and revolted by the idea of a man making love to a fetish object, or copulating with another male, yet they happily accept parental mal-imprinting where a human adult is fondling a pet lap-dog or feeding a baby monkey from a bottle. But why do they make the distinction? Biologically speaking, there is virtually no difference between the two activities. They both involve mal-imprinting and they are both aberrations of normal human relationships. But although, in the biological sense, they must both be classed as abnormalities, neither of them causes any harm to bystanders, to individuals outside the relationships. We may feel that it would be more gratifying for the fetishist or the childless animal lover if they could enjoy the rewards of a full family life, but it is their loss, not ours, and we have no cause to be hostile to either of them.

We have to face the fact that, living in a human zoo, we are inevitably going to suffer from many abnormal relationships. We are bound to be exposed in unusual ways to unusual stimuli. Our nervous systems are not equipped to deal with this and our patterns of response will sometimes misfire. Like the experimental or zoo animals, we may find ourselves fixated with strange and sometimes damaging bonds, or we may suffer from serious bond confusion. It can happen to any of us, at any time. It is merely another of the hazards of existing as an inmate of a human zoo. We are all potential victims, and the most appropriate reaction, when we come across it in someone else, is sympathy rather than cold intolerance.

CHAPTER SIX

THE STIMULUS STRUGGLE

WHEN a man is reaching retiring age he often dreams of sitting quietly in the sun. By relaxing and 'taking it easy' he hopes to stretch out an enjoyable old age. If he manages to fulfil his sun-sit dream, one thing is certain: he will not lengthen his life, he will shorten it. The reason is simple—he will have given up the Stimulus Struggle. In the human zoo this is something we are all engaged in during our lives and if we abandon it, or tackle it badly, we are in serious trouble.

The object of the struggle is to obtain the optimum amount of stimulation from the environment. This does not mean the maximum amount. It is possible to be over-stimulated as well as under-stimulated. The optimum (or happy medium) lies somewhere between these two extremes. It is like adjusting the volume of music coming from a radio: too low and it makes no impact, too high and it causes pain. At some point between the two there is the ideal level, and it is obtaining this level in relation to our whole existence that is the goal of the Stimulus Struggle.

For the super-tribesman this is not easy. It is as if he were surrounded by hundreds of behaviour 'radios', some whispering and others blaring away. If, in extreme situations, they are all whispering, or monotonously repeating the same sounds over and over again, he will suffer from acute boredom. If they are all blaring, he will experience severe stress.

Our early tribal ancestor did not find this such a difficult problem. The demands of survival kept him busy. It required all his time and energy to stay alive, to find food and water, to defend his territory, to avoid his enemies, to breed and rear his young and to construct and maintain his shelter. Even when times were exceptionally bad, the challenges were at least comparatively straightforward. He can never have been subjected to the intricate and complex frustrations and conflicts that have become so typical of super-tribal existence. Nor is he likely to have suffered unduly from the boredom of gross under-stimulation that, paradoxically, super-tribal life can also impose. The advanced forms of the Stimulus Struggle are therefore a speciality of the urban animal. We do not find them among wild animals or 'wild' men in their natural environments. We do, however, find them in both urban men and in a particular kind of urban animal—the zoo inmate.

Like the human zoo, the animal zoo provides its occupants with the security of regular food and water, protection from the elements and freedom from natural predators. It looks after their hygiene and their health. It may also, in certain cases, put them under severe strain. In this highly artificial condition, zoo animals, too, are forced to switch from the struggle for survival to the Stimulus Struggle. When there is too little input from the world around them, they have to contrive ways of increasing it. Occasionally, when there is too much (as in the panic of a freshly caught animal), they have to try and damp it down.

The problem is more serious for some species than for others. From this point of view there are two basic kinds of animals: the specialists and the opportunists. The specialists are those which have evolved one supreme survival device on which they depend for their very existence, and which dominates their lives. Such creatures are the anteaters, the koalas, the giant pandas, the snakes and the eagles. So long as ant-eaters have their ants, koalas have their eucalyptus leaves, pandas have their bamboo shoots, and snakes and eagles have their prey, they can relax. They have perfected their diet specializations to such a pitch that, providing their particular requirements are met, they can accept a lazy and otherwise unstimulating pattern of life. Eagles, for instance, will thrive in a small empty cage for over forty years without so much as biting their claws, providing, of course, they can sink them daily into a freshly killed rabbit.

The opportunists are not so fortunate. They are the species—such as dogs and wolves, raccoons and coatis, and monkeys and apes—that have evolved no single, specialized survival device. They are jacks-of-all-trades, always on the look-out for any small advantage the environment has to offer. In the wild, they never stop exploring and investigating. Anything and everything is examined in case it may add yet another string to the bow of survival. They cannot afford to relax for very long and evolution has made sure that they do not. They have evolved nervous systems that abhor inactivity, that keep them constantly on the go. Of all species, it is man himself who is the supreme opportunist. Like the others, he is intensely exploratory. Like them, he has a biologically built-in demand for a high stimulus input from his environment.

In a zoo (or a city) it is clearly these opportunist species that will suffer most from the artificiality of the situation. Even if they are provided with perfectly balanced diets and are immaculately sheltered and protected, they will become bored and listless and eventually neurotic. The more we have come to understand the natural behaviour of such animals, the more obvious it has become, for example, that zoo monkeys are little more than distorted caricatures of their wild counterparts.

But opportunist animals do not give up easily. They react to the unpleasant situation with remarkable ingenuity. So, too, do the inmates of the human zoo. If we compare the animal zoo reactions with those we find in the human zoo, it will serve to bring home to us the striking parallels that exist between these two highly artificial environments.

The Stimulus Struggle operates on six basic principles, and it will help if we look at them one by one, examining in each case first the animal zoo and then the human zoo. The principles are these:

1. If stimulation is too weak, you may increase your behaviour output *by creating unnecessary problems which you can then solve.*

We have all heard of labour-saving devices, but this principle is concerned with labour-wasting devices. The Stimulus Struggler deliberately makes work for himself by elaborating patterns that could otherwise be performed more simply, or that need no longer be performed at all.

In its zoo cage, a wild cat may be seen to throw a dead bird or a dead rat up into the air and then leap after it and pounce on it. By throwing the prey, the cat can put movement and therefore 'life' back into it, giving itself the chance to perform a 'kill'. In the same way, a captive mongoose can be seen 'shaking to death' a piece of meat.

Observations of this kind extend to domestic animals as well. A pet dog, pampered and well fed, will drop a ball or a stick at its master's feet and wait patiently for the object to be thrown. Once it is moving through the air or across the ground, it becomes 'prey' and can be chased after, caught, 'killed' and brought back again for a repeat performance. The domestic dog may not be hungry for food, but it is hungry for stimulation.

In its own way, a caged raccoon is equally ingenious. If there is no food to search for in a near-by stream, the animal will search for it anyway, even if there is no stream. It takes its food to its water-dish, drops it in, loses it and then searches for it. When it finds it, it scrabbles with it in the water before eating it. Sometimes it even destroys it by this process, pieces of bread becoming a hopeless mush. But no matter, the frustrated food-searching urge has been satisfied. This, incidentally, is the origin of the long-standing myth that raccoons wash their food.

There is a large rodent, looking like a guinea-pig on stilts, called an agouti. In the wild it peels certain vegetables before eating them. It holds the vegetables in its front feet and pares them with its teeth as we might pare an orange. Only when it has completely skinned the object does it start to eat. In captivity, this peeling urge refuses to be frustrated. If a perfectly clean apple or potato is given to an agouti, the animal still peels it fastidiously and, after eating it, devours the peel as well. It even attempts to 'peel' a piece of bread.

Turning to the human zoo, the picture is strikingly similar. When we are born into a modern super-tribe, we are thrust into a world where human brilliance has already solved most of the basic survival problems. Just like the zoo animals, we find that our environment emanates security. Most of us have to do a certain amount of work, but thanks to technical developments, there is plenty of time left over for participating in the Stimulus Struggle. We are no longer totally absorbed in the problems of finding food and shelter, rearing our offspring, defending our territories or avoiding our enemies. If, against this, you argue that you never stop working, then you must ask yourself a key question: could you do less work and still survive? The answer in many cases would have to be 'yes'. Working is the modern super-tribesman's equivalent of hunting for food and, like the animal zoo inmates, he frequently performs thc pattern much more elaborately than is strictly necessary. He creates problems for himself.

Only those sectors of the super-tribe that are enduring what we would call severe hardship are working totally for survival. Even they, however, will be forced to indulge in the Stimulus Struggle when they can spare a moment, for the following special reason: The primitive, hunting tribesman may have been a 'survival-worker', but his tasks were varied and absorbing. The unfortunate subordinate super-tribesman who is a 'survival-worker' is not so well off. Thanks to the division of labour and industrialization, he is driven to carry out intensely dull and repetitive work—the same routine thing day after day, year after year—making a mockery of the giant brain housed inside his skull. When he does get a few moments to himself, he needs to indulge in the Stimulus Struggle as much as anyone else in our modern world, for the problem of stimulation is concerned with variety as well as amount, with quality as well as quantity.

For the others, as I have said, much of the activity is work for work's sake and, if it is exciting enough, the struggler—a businessman, for instance—may find that he has scored so many points during his working day that, in his spare time, he can allow himself to relax and indulge in the mildest of activities. He might doze at his fireside with a soothing drink, or dine out at a quiet restaurant. If he dances when he dines, it is worth observing how he does it. The point is that our survival-worker may also go dancing in the evening. At first sight there appears to be a contradiction here, but closer examination reveals that there is a world of difference between the two kinds of dancing. Big-businessmen do not go in for strenuous competitive ballroom dancing, or wild abandoned folk-dancing. Their clumsy shuffling on the night-club floor (the small size of which has been tailored to their low-stimulus demands) is far from being competitive or wild. The unskilled workman is likely to become a skilled dancer; the skilled businessman is likely to be an unskilled dancer. In both cases the individual achieves a balance which is, of course, the goal of the Stimulus Struggle.

In over-simplifying to make this point I have made the difference between the two types sound too much like a class distinction, which it is not. There are plenty of bored businessmen, suffering from repetitive office tasks that are almost as monotonous as packing boxes at a factory bench. They too will have to seek more stimulating forms of recreation in their spare time. Also, there are many simple labouring jobs where the work is rich and varied. The more fortunate labourer, in the evening, is more like the successful businessman, relaxing with a quiet drink and a chat.

The under-stimulated housewife is another interesting phenomenon. Surrounded by her modern labour-saving devices, she has to invent labour-wasting devices to occupy her time. This is not as futile as it sounds. She can at least *choose* her activities: therein lies the whole advantage of super-tribal living. In primitive tribal life there was no choice. Survival made its own demands. You had to do this, and this, and this, or die. Now you can do this, or that, or the other—anything you like, so long as you realize that you have to do *something*, or break the golden rules of the Stimulus Struggle. And so the housewife, her washing spinning automatically away in the kitchen, must busy herself with something else. The possibilities are endless and the game can be a most attractive one. It can also go astray. Every so often it suddenly seems to the under-stimulated player that the compensating activity he or she is pursuing so relentlessly is really rather meaningless. What *is* the point of rearranging the furniture, or collecting postage stamps, or entering the dog for another dog-show? What does it prove? What does it achieve? This is one of the dangers of the Stimulus Struggle. Substitutes for real survival activity remain substitutes, no matter how you look at them. Disillusionment can easily set in, and then it has to be dealt with.

There are several solutions. One is a rather drastic one. It is a variation of the Stimulus Struggle called Tempting Survival. The disillusioned teenager, instead of throwing a ball on a playing field, can throw it through a plate-glass window. The disillusioned housewife, instead of stroking the dog, can stroke the milkman. The disillusioned businessman, instead of stripping down the engine of his car, can strip down his secretary. The ramifications of this manoeuvre are dramatic. In no time at all the individual is involved in the true survival struggle of fighting for his social life. During such phases there is a characteristic loss of interest in furniture rearranging and postage-stamp collecting. After the chaos has died down, the old substitute activities suddenly seem more appealing again.

A less drastic variant is Tempting Survival By Proxy. One form this takes consists of meddling in other people's emotional lives and creating for them the sort of chaos that you would otherwise have to go through yourself. This is the malicious gossip principle: it is extremely popular because it is so much safer than direct action. The worst that can happen is that you lose some of your friends. If it is operated skilfully enough, the reverse may occur: they may become substantially *more* friendly. If your machinations have succeeded in breaking up their lives, they may have a greater need of your friendship than ever before. So, providing you are not caught out, this variation can have a double benefit: the vicarious thrill of watching their survival drama, and the subsequent increase in their friendliness.

A second form of Tempting Survival By Proxy is less damaging. It consists of identifying yourself with the survival drama of fictional characters in books, films, plays and on television. This is even more popular, and a giant industry has grown up to meet the enormous demands it creates. It is not only harmless and safe, but it also has the distinction of being remarkably inexpensive. The straight game of Tempting Survival can end up costing thousands, but this variant, for no more than a few shillings, can permit the Stimulus Struggler to indulge in seduction, rape, adultery, starvation, murder and pillage, without so much as leaving the comfort of his chair.

2. If stimulation is too weak, you may increase your behaviour output *by over-reacting to a normal stimulus.*

This is the over-indulgence principle of the Stimulus Struggle. Instead of setting up a problem to which you then have to find a solution, as in the last case, you simply go on and on reacting to a stimulus that is already to hand, although it no longer excites you in its original role. It has become merely an occupational device.

In zoos where the public are permitted to feed the animals, certain bored species with nothing else to do will continue to eat until they become grossly over-weight. They will have already eaten their complete zoo diet and are no longer hungry, but idle nibbling is better than doing nothing. They get fatter and fatter, or become sick, or both. Goats eat mountains of ice-cream cartons, paper, almost anything they are offered. Ostriches even consume sharp metal objects. A classic case concerns a female elephant.

She was observed closely for a single typical zoo day and during that period (in addition to her normal, nutritionally adequate zoo diet) she devoured the following objects offered to her by the public: 1,706 peanuts, 1,330 sweets, 1,089 pieces of bread, 811 biscuits, 198 segments of orange, 17 apples, 16 pieces of paper, 7 ice-creams, 1 hamburger, 1 boot-lace and 1 lady's white leather glove. There are cases on records of zoo bears dying of suffocation by the enormous pressure of food in their stomachs. Such are the sacrifices made to the Stimulus Struggle.

One of the strangest examples of this phenomenon concerns a large male gorilla which regularly ate, regurgitated and then re-ate his food, performing his own version of a Roman banquet. This process was taken a stage further by a sloth bear which was frequently observed to regurgitate its food more than a hundred times, each time eating it up again with the gurgling and sucking sounds typical of its species.

If the possibilities of over-indulging in feeding behaviour are limited and there is nothing else to do, an animal can always clean itself excessively, extending the performance until long after its feathers or its fur are perfectly cleansed and groomed. This, too, can lead to trouble. I recall one sulphur-creasted cockatoo that had only a single feather left, a long yellow crest-plume, the rest of its body being as naked as an oven-chicken's. That was an extreme case, but not an isolated one. Mammals can scratch and lick bare patches until sores develop and set up their own vicious circle of irritation and picking.

For the human Stimulus Struggler, the unpleasant forms that this principle takes are well known. In infancy there is the example of prolonged thumb-sucking, which results from too little contact and inter-action with the mother. As we grow older we can indulge in occupational eating, nibbling aimlessly away at chocolates and biscuits to pass the time, and getting fatter and fatter as a result, like the zoo bears. Or we can groom ourselves into trouble, like the cockatoo. For us it will probably take the form of nail-biting or scab-picking. Occupational drinking, if the drinks are long and sweet, can lead again to fatness; if short and alcoholic they can lead to addiction and possibly liver damage. Smoking can be another time-killer and this, too, has its dangers.

Clearly there are pitfalls if the Stimulus Struggle is tackled badly. The snag with these over-indulgence time-killers is that they are so limited that they make development impossible. All one can do with them is repeat them over and over again, to stretch them out. To be effective in a major way they must be indulged in for long periods and that means trouble. Harmless enough in the ordinary course of events, as minor time-killers, they become damaging when carried to excess.

3. If stimulation is too weak, you may increase your behaviour output by *inventing novel activities*.

This is the creativity principle. If familiar patterns are too dull, the intelligent zoo animal must invent new ones. Captive chimpanzees, for instance, will contrive to introduce novelty into their environment by exploring the possibilities of new forms of locomotion, rolling over and over, dragging their feet along and performing a variety of gymnastic patterns. If they can find a small piece of string, they will thread it through the cage roof, hang on to both ends with their teeth or their hands and spin it round in the air, suspended like circus acrobats.

Many zoo animals use visitors to relieve the boredom. If they ignore the people who walk by their cages they are liable to be ignored in return, but if they stimulate them in some way, then the visitors will stimulate them back. It is surprising what you can get zoo visitors to do, if you are an ingenious zoo animal. If you are a chimpanzee or an orang-utan and you spit at them, they scream and rush wildly about. It helps to pass the day. If you are an elephant, you can flick spittle at them with the tip of your trunk. If you are a walrus, you can splash water over them with your flipper. If you are a magpie or a parrot, you can entice them with ruffled head feathers to preen you and then nip their fingers with your beak.

One particular male lion perfected his audience-manipulation in a remarkable way. His usual method of urination (as with tom-cats) was to squirt a jet of urine horizontally backwards at a vertical landmark, depositing his personal scent upon it. When he did this against one of the vertical bars of his cage-front he found that the spray reached his visitors and created an interesting reaction. They leapt back, shouting. As time passed, he not only improved his aim, but also added a new trick. After the first spraying, when the front row of his audience had retreated, the second row quickly took its place to get a better look. Instead of loosing his jet in one stream, he saved some of it for a second spraying and in this way managed to excite the new front row as well.

Food-begging (as distinct from food-nibbling) is a less drastic measure, but equally rewarding, and is practised by a wide variety of species. All that is necessary is to invent some peculiar action or posture that appeals to the passers-by and makes them believe that you are hungry. Monkeys and apes find that an out-stretched palm is adequate, but bears have proved more inventive. Each has its own speciality: one will stand on its hind legs and wave a paw; another will sit on its rump in a curved posture, clasping its hind paws with its front feet; another will sit up and hook one of its front paws on to the lower jaw of its open mouth; another will stand up and nod or make come-hither movements with its head. It is amazing how easy it is to train zoo visitors to react to these displays if you are an intelligent zoo bear. The trouble is that in order to keep the visitors' interest, you have to reward them every so often by eating the objects they throw at you. If you fail to comply, they soon move away and the exciting stimulation of the social interaction you have invented is lost. The result of this we have already observed: you have to switch to the less satisfactory 'over-indulgence principle' and you get fat and sick.

The essential point about these zoo gymnastics and begging routines is that the motor patterns involved are not found in nature. They are inventions geared to the special conditions of captivity.

In the human zoo this creativity principle is carried to impressive extremes. I have already pointed out that disillusionment can set in when the survival-substitute activities of the Stimulus Struggle begin to seem pointless, often because the activities chosen are rather limited in their scope. In avoiding these limitations, men have sought for more and more complex forms of expression, forms which become so absorbing that they carry the individual on to such high planes of experience that the rewards are endless. Here we move from the realms of occupational trivia to the exciting worlds of the fine arts, philosophy and the pure sciences. These have the great value that they not only effectively combat under-stimulation, but also at the same time make maximum use of man's most spectacular physical property—his gigantic brain.

Because of the vast importance these activities have assumed in our civilizations, we tend to forget that they are in a sense no more than devices of the Stimulus Struggle. Like hide-and-seek or chess, they help to pass the time between the cradle and the grave, for those who are lucky enough not to be totally bound up in the struggle for crude survival. I say lucky, because, as I mentioned earlier, the great advantage of the super-tribal condition is that we are comparatively free to choose the forms that our activities take, and when the human brain can devise such beautiful pursuits as these, we must count ourselves fortunate to be among the Stimulus Strugglers, rather than the strugglers for survival. This is man the inventor playing for all he is worth. When we study the researches of science, listen to symphonies, read poetry, watch ballets or look at paintings, we can only marvel at the lengths to which mankind has pushed the Stimulus Struggle and the incredible sensitivity with which he has tackled it.

4. If stimulation is too weak, you may increase your behaviour output by *performing normal responses to sub-normal stimuli*

This is the overflow principle. If the internal urge to perform some activity becomes too great, it can 'overflow' in the absence of the external objects that normally provoke it.

Objects which in the wild state would never rate a reaction are given the full treatment in the bleak zoo environment. With monkeys this may take the form of coprophagy: if there is no food to chew, then faeces will do. If there is no territory to patrol, then stunted cage-pacing will do. The animal ambles back and forth, back and forth, until it has worn a track by its rhythmic, sterile pacing. Again, it is better than nothing.

In the absence of a suitable mate, a zoo animal may

attempt to copulate with virtually anything that is available. A solitary hyena, for example, managed to mate with its circular food-dish, tipping it up on its side and rolling it back and forth beneath its body so that it pressed rhythmically against its penis. A male raccoon living alone used its bed as a mate. It could be seen to gather up a tight bundle of straw, clasp it beneath its body and then make pelvic thrusts into it. Sometimes, when an animal is kept with another of a different species, the alien companion can be used as a mate substitute. A male brush-tailed porcupine living with a tree-porcupine repeatedly tried to mount it. The two species are not closely related and the arrangement of the spines differs markedly, with the result that it was an extremely painful affair for the frustrated male. In another cage a little squirrel monkey was housed with a large kangaroo-shaped rodent called a springhaas, which was about ten times its size. Undaunted, the diminutive monkey used to leap on to the sleeping rodent's back and attempt to copulate. The result of its desperate frustrations were reported in the local press, but totally misunderstood. It was recorded as having indulged in a charming game, 'riding on the big animal's back like a little furry jockey'.

The sexual examples are reminiscent of fetishism, but must not be confused with it. In the case of 'overflow activities', as soon as the natural stimulus is introduced into the environment, the animal reverts to normality. In the instances I have mentioned, the males immediately switched their attentions to females of their own species when these became available. They were not 'hooked' on their female-substitutes, like the true fetishists I discussed in the last chapter.

An unusual mutual overflow activity occurred when a female sloth and a small douroucouli monkey were housed together. In nature this monkey makes a snug den for itself inside a hollow tree, where it sleeps during the day. The female sloth, had she given birth in the wild, would have carried her offspring on her body for a considerable period. In the zoo, the monkey lacked a warm, snug bed and the sloth lacked an offspring. The problem was neatly solved for both of them by the simple expedient of the monkey sleeping clamped tightly on to the sloth's body.

The operation of this fourth principle of the Stimulus Struggle is not so much a case of any-port-in-a-storm as any-port-when-becalmed and, despite the many winds that blow through the human zoo, the human animal frequently finds itself in this sort of situation. The emotional patterns of the super-tribesman are constantly being blocked for one reason or another. In the midst of material plenty there is much behavioural deprivation. Then he, like the zoo animals, is driven to respond to sub-normal stimuli, no matter how inferior these may be.

In the sexual sphere, man is better equipped than most species to solve the absent-mate problem by masturbating, and this is the most common human solution. Despite this, zoophilia, or the act of copulation performed between a human being and some other animal species, does occur from time to time. It is rare, but less rare than most people imagine. A recent American survey revealed that in that country, among boys raised on farms, about 17 per cent experience orgasm as a result of 'animal contacts' at least once during their lives. There are many more that indulge in milder forms of sexual interaction with farm animals, and in certain districts the total figure has been put as high as 65 per cent of farm boys. The animals favoured are usually calves, donkeys and sheep, and occasionally some of the larger birds such as geese, ducks and chickens.

Zoophilic activities are much rarer among human females. Out of nearly six thousand American women only twenty-five had experienced orgasm as a result of stimulation by another animal species, usually a dog.

To most people such activities seem bizarre and revolting. The fact that they occur at all reveals the extraordinary lengths to which Stimulus Strugglers will go in avoiding inactivity. The parallel with the zoo world is inescapable.

Other forms of sexual behaviour, such as certain cases of 'better-than-nothing' homosexuality, also fall into this category. In the absence of normal stimulation, the subnormal object becomes adequate. Starving men will chew wood and other nutritionally worthless objects, rather than chew nothing. Aggressive individuals with no enemies to attack will violently smash inanimate objects or mutilate their own bodies.

5. If stimulation is too weak, you may increase your behaviour output *by artificially magnifying selected stimuli.*

This principle concerns the creation of 'super-normal stimuli'. It operates on the simple premise that if natural, normal stimuli produce normal responses, then super-normal stimuli should produce super-normal responses. This idea has been put to great use in the human zoo, but it is rare in the animal zoo. Students of animal behaviour have devised a number of super-normal stimuli for experimental animals, but the accidental occurrence of the phenomenon is limited to only a few examples, one of which I will describe in detail.

It stems from my own research. For some time I had been keeping a mixed collection of birds in a large aviary on the roof of a research department. At one point they became troubled by nocturnal visits from a predatory owl which attempted to attack them through the wire of the aviary.

Investigating the problem led me to make a number of dusk watches. The owl never came while I was there, and in fact was never heard of again, but although I drew a blank in that respect, what I did see was some very strange behaviour going on inside the aviary itself.

Among the birds were some doves and some small finches called Java sparrows. These finches normally roost together, pressed closely up against one another on a branch. To my surprise, the finches in the aviary were ignoring one another, favouring the doves instead as roosting companions. Each dove had a tiny finch pressed tightly up against its plump body. The small birds were snuggling down contentedly for the night, and the doves, although somewhat startled at first by their strange sleeping partners, were too drowsy to do anything about them, and eventually they too settled down for the night's sleep.

I was completely at a loss to explain this peculiar pattern of behaviour. The two species had not been reared together, so there could be no question of mal-imprinting. The finches had not even been bred in captivity. They should, by all the rules, have roosted with other members of their own species. There was another problem. Why, out of all the other species in the aviary, did they choose the doves to sleep with?

Returning to my roost-time vigil on subsequent nights, I was able to observe even more curious behaviour. Before going to sleep, the tiny finches often preened their doves, again an action which under normal circumstances they would only direct towards one of their own kind. Stranger still, they began to play leapfrog over the backs of their huge companions. A finch would leap on to the back of its dove, then off again at the other side; then back again, and so on. The ultimate oddity came when I saw one of the small birds push up underneath the body of its dove and shove itself between the big bird's legs. The sleepy dove stretched high on its legs and stared down at the struggling form beneath its rounded breast. Once in position, the finch settled down and the dove subsided on to it. There they sat, with the finch's pink beak protruding from the bottom of the dove's chest.

Somehow I had to find an explanation for this extraordinary relationship. There was nothing odd about the doves, except perhaps their remarkable tolerance. It was the finches that demanded further study. I found that they had a special signal at roosting time that indicated to other members of their species that they were ready to go to sleep. When they were active they kept their distance from one another, but when it was time to clump together for the night, one finch, presumably the sleepiest, would fluff out its feathers and squat low on its perch. This was the signal to other members of its group that they could join it without being repulsed. A second finch would fly in and squat up against the first one, fluffing its feathers out as it did so; then a third, a fourth and so on, until a row of roosting birds had been formed. Late-comers would often hop along the backs of the row and squeeze down into a warmer and more favourable position in the middle. Here were all the clues I needed.

The combined fluffing-and-squatting action made the finches look bigger and more spherical than when they were actively moving about. This was the key signal, saying, 'come roost with me.' A roosting dove was even bigger and more spherical, and therefore could not help sending out a much more powerful version of the same signal. Furthermore, unlike the other species in the aviary, the doves had the same greyish colour as the little finches. As they were so big, rounded and grey, they gave out a *super-normal signal* to the finches which the small birds simply could not resist.

Being innately programmed to this combination of size, shape and colour, the finches automatically responded to the doves as super-normal stimuli for roosting, preferring them to their own species. The snag was that the doves did not form rows. A finch clumping with one found itself at the end of a 'row', jumped on to the dove's back, failed to find the middle of the 'row' and jumped off the other side. The dove was so big that it must have seemed like a whole row of finches, so the small bird tried again, but still without success. With great persistence, the finch eventually tried pushing up from underneath the dove and at last found a snug position in the 'middle of the row', between the bigger bird's legs.

As I said earlier, this is one of the few known instances of a non-human super-normal stimulus occurring without a deliberate experiment being carried out. Other, better known examples have always involved the use of an experimental dummy. Oystercatchers, for instance, are ground-nesting birds. If one of their eggs rolls out of the nest, it is pulled back in with a special action of the beak. If dummy eggs are placed near the nest, the birds will pull these in too. If offered dummy eggs of different sizes, they always prefer the biggest one. They will, in fact, try to heave in eggs many times the size of their own real eggs. Again, they cannot help reacting to a super-normal stimulus.

Herring gull chicks, when they beg for food from their parents, peck at a bright red spot that is situated near the tip of the adult birds' bills. The parents respond to this pecking by regurgitating fish for their young. The red spot is the vital signal. It was discovered that the chicks would even peck at flat cardboard models of their parents' heads. By a series of tests it was found that the other details of the adult head were unimportant. The chicks would peck at a red spot by itself. Furthermore, if they were offered a stick with three red spots on it, they would actually peck *more* at that than at a complete and realistic model of their parents. Again, the stick with the three red spots was a super-normal stimulus.

There are other examples, but these will suffice. Clearly, it is possible to improve on nature, a fact which some have found distasteful. But the reason is simple: each animal is a complex system of compromises. The conflicting demands of survival pull it in different directions. If, for example, it is too brightly coloured, it will be detected by its predators. If it is too drably coloured, it will be unable to attract a mate, and so on. Only when the pressures of survival are artificially reduced will this system of compromises be relaxed. Domesticated animals, for instance, are protected by man and no longer need fear their predators. Without risk, their dull colours can be replaced by pure white, gaudy piebalds and other vivid patterns. But if they were turned loose again in their natural habitat, they would be so conspicuous that they would quickly fall prey to their natural enemies.

Like his domesticated animals, super-tribal man can also afford to ignore the survival restrictions of natural stimuli. He can manipulate stimuli, exaggerate them and distort them to his heart's content. By increasing their strength artificially—by creating super-normal stimuli—he can give an enormous boost to his responsiveness. In his super-tribal world he is like an oystercatcher surrounded by giant eggs.

Everywhere you look you will find evidence of some kind of super-normal stimulation. We like the colours of flowers, so we breed bigger and brighter ones. We like the rhythm of human locomotion, so we develop gymnastics. We like the taste of food, so we make it spicier and tastier. We like certain scents, so we manufacture strong perfumes. We like a comfortable surface to sleep on, so we construct super-normal beds with springs and mattresses.

We can start by examining our appearance—our clothes and our cosmetics. Many male costumes include padding of the shoulders. At puberty there is a marked difference in the growth rate of the shoulders in males and females, those of boys becoming broader than those of girls. This is a natural, biological signal of adult masculinity. Padding the shoulders adds a super-normal quality to this masculinity and it is not surprising that the most exaggerated trend occurs in that most masculine of spheres, the military, where stiff epaulets are added to further increase the effect. A rise in body height is also an adult feature, especially in males, and many an aggressive costume is crowned by some form of tall head-gear, creating the impression of super-normal height. We would no doubt wear stilts, too, if they were not so cumbersome.

If males wish to appear super-normally young, they can wear toupees to cover their bald heads, false teeth to fill their ageing mouths and corsets to hold in their sagging bellies. Young executives, who wish to appear super-normally old, have been known to indulge in artificial greying of their juvenile hair.

The adolescent female of our species undergoes a swelling of the breasts and widening of the hips that mark her out as a developing sexual adult. She can strengthen her sexual signals by exaggerating these features. She can raise, pad, point or inflate her breasts in a variety of ways. By tightening her waist she can throw into contrast the width of her hips. She can also pad out her buttocks and her hips, a trend that found its most super-normal development in the periods of bustles and crinolines.

Another growth change that accompanies the maturation of the female is the lengthening of the legs in relation to the rest of the body. Long legs can therefore come to equal sexuality and exceptionally lengthy legs become sexually appealing. They cannot, of course, become super-normal stimuli themselves, being natural objects (although high heels will help a little), but artificial lengthening can occur in erotic drawings and paintings of females. Measurements of drawings of 'pin-ups' reveal that the girls are usually portrayed with unnaturally long legs, sometimes almost one and a half times as long as the legs of the models on which they are based. The recent fashion for very short skirts owes its sexual appeal not simply to the exposure of bare flesh, but also to the impression of longer legs it gives when contrasted with the earlier longer-skirted styles.

A glittering array of super-normal stimuli can be found in the world of female cosmetics. A clear, unblemished skin is universally attractive sexually. Its smoothness can be exaggerated by powders and creams. At times when it has been important to show that a female did not have to toil in the sun, her cosmetics aided her by creating a super-normal whiteness for her visible skin. When conditions changed and it became important for her to reveal that she could afford the *leisure* to lie in the sun, then tanning of the skin became an asset. Once again her cosmetics were there to provide her with super-normal browning. At other periods, in the past, it was important that she displayed her healthiness, and the super-normal flush of rouge was added. Another feature of her skin is that it is less hairy than that of the adult male. Here again, a super-normal effect can be achieved by various forms of depilation, the tiny hairs being shaved or stripped from the legs, or painfully plucked from the face. The eye-brows of the male tend to be bushier than those of the female, so super-normal femininity can be obtained by plucking here, too. Add to all this her super-normal eye make-up, lipstick, nail-varnish, perfume and occasionally even nipple-rouge, and it is easy to see how hard we work the super-normal principle of the Stimulus Struggle.

We have already observed in a previous chapter the lengths to which the male penis has gone in becoming a super-normal phallic symbol. In ordinary clothing it has not fared so well, except for a brief moment of glory during the epoch of the codpiece. Today we are left with little more than the super-normal pubic tuft of the Scotsman's sporran.

The strange world of aphrodisiacs is entirely devoted to the subject of super-normal sexual stimuli. For many centuries and in many cultures, ageing human males have attempted to boost their waning sexual responses by means of artificial aids. A dictionary of aphrodisiacs lists over nine hundred items, including such delightful potions as angel water, camel's hump, crocodile dung, deer sperm, goose tongues, hare soup, lion's fat, necks of snails and swan's genitals. Doubtless many of these aids proved successful, not because of their chemical properties, but because of the inflated prices paid for them. In the eastern world, powdered rhino horn has been so highly valued as a super-normal sexual stimulus that certain species of rhinoceros have nearly become extinct. Not all aphrodisiacs were swallowed. Some were rubbed on, others smoked, sniffed or worn on the body. Everything from aromatic baths to scented snuff seems to have been pressed into service in the frantic search for stronger and more violent stimulation.

The modern pharmacy is less sexually orientated, but it is bulging with super-normal stimuli of many kinds. There are sleeping pills to produce super-normal sleep, pep pills to produce super-normal alertness, laxatives to produce super-normal defecation, toilet preparations to produce super-normal body-cleaning and toothpaste to produce a super-normal smile. Thanks to man's ingenuity there is hardly any natural activity which cannot be provided with some form of artificial boost.

The world of commercial advertising is a seething mass of super-normal stimuli, each trying to out-pace the others. With competing firms marketing almost identical products, the super-normal Stimulus Struggle has become big business. Each product has to be presented in a more stimulating form than its rivals. This requires endless attention to subtleties of shape, texture, pattern and colour.

An essential feature of a super-normal stimulus is that it need not involve an exaggeration of *all* the elements of the natural stimulus on which it is based. The oystercatcher responded to a dummy egg that was super-normal in only one respect—its size. In shape, colour and texture it was similar to a normal egg. The experiment with the gull chicks went one step further. There, the vital red spots were exaggerated and, in addition, the other features of the parent figure, the unimportant ones, were eliminated. A double process was therefore taking place: magnification of the essential stimuli and, at the same time, elimination of the inessential ones. In the experiment this was done merely to demonstrate that the red spots alone were sufficient to trigger the reaction. Nevertheless, taking this step must also have helped in focusing more attention on the red spots by removing irrelevancies. With many human super-normal stimuli this dual process has been employed with great effect. It can be expressed as an additional, subsidiary principle for the Stimulus Struggle:

This states that when selected stimuli are magnified artificially to become super-normal stimuli, the effect can be further enhanced by reducing other (non-selected or irrelevant) stimuli. By simultaneously creating sub-normal stimuli in this way, the super-normal stimuli appear relatively stronger. This is the principle of *stimulis extremism.*

If we wish to be entertained by books, plays, films or songs, we automatically subject ourselves to this procedure. It is the very essence of the process we call dramatization. Everyday actions performed as they happen in real life would not be exciting enough. They have to be exaggerated. The operation of the stimulus extremism principle ensures that irrelevant detail is suppressed and relevant detail is heightened and made more extravagant. Even in the most realistic schools of acting, or, for that matter, in non-fiction writing and documentary filming, the negative process still operates. Irrelevancies are pared away, thus producing an indirect form of exaggeration. In the more stylized performances, such as opera and melodrama, the direct forms of exaggeration are more important and it is remarkable to see how far the voices, the costumes, the gestures, the actions and the plot can stray away from reality and yet still make a powerful impact on the human brain. If this seems strange, it is worth recalling the case of the experimental birds. The gull chicks were prepared to respond to a substitute for their parents that consisted of something as remote from an adult gull as a stick with three red spots on it. Our reactions to the highly stylized rituals of an opera are no more outlandish.

Children's toys, dolls and puppets illustrate the same principle very vividly. A rag doll's face, for example, has certain important features magnified and others omitted. The eyes become huge black spots, while the eyebrows disappear. The mouth is shown in a vast grin, while the nose is reduced to two small dots. Enter a toyshop and you enter a world of contrasting super-normal and sub-normal stimuli. Only the toys for the older children become less contrasted and more realistic.

The same is true of the children's own drawings. In portrayals of the human body, those features that are important to them are enlarged; those that are unimportant are reduced or omitted. Usually the head, eyes and mouth receive the most disproportionate magnification. These are the parts of the body that have most meaning for a young child, because they form the area of visual expression and communication. The external ears of our species are inexpressive and comparatively unimportant and they are therefore frequently left out altogether.

Visual extremism of this kind is also prevalent in the arts of primitive peoples. The size of heads, eyes and mouths is usually super-normal in relation to the dimensions of the body and, as with children's drawings, other features are reduced. The stimuli selected for magnification do vary from case to case, however. If a figure is shown running, then its legs become super-normally large. If a figure is simply standing and is doing nothing with either its arms or its legs, they may become mere stumps or disappear altogether. If a prehistoric figurine is concerned with representing fertility, its reproductive features may become super-normalized to the exclusion of all else. Such a figure may boast a huge pregnant belly, enormous protruding buttocks, wide hips and vast breasts, but have no legs, arms, neck or head.

Graphic manipulations of subject-matter in this way have often been referred to as the creation of ugly deformities, as if the beauty of the human form were somehow being subjected to malicious damage and insult. The irony is that if such critics examined their own bodily adornments they would find that their own appearance was not exactly 'as nature intended'. Like the children and the primitive artists they are no doubt laden with 'deforming' super-normal and sub-normal elements.

The fascination of stimulus extremism in the arts lies in the way these exaggerations vary from case to case and place to place, and in the way the modifications develop new forms of harmony and balance. In the modern world, animated cartoon films have become major purveyors of this type of visual exaggeration, and a specialized form of it is to be found in the art of caricature. The expert caricaturist picks out the naturally exaggerated features of his victim's face and deftly super-normalizes these already existing exaggerations. At the same time he reduces the more inconspicuous features. The magnification of a large nose, for example, can become so extreme that it ends up with its dimensions doubled or even tripled, without rendering the face unrecognizable. Indeed, it makes it even more recognizable. The point is that we all identify individual faces by comparing them in our minds with an idealized 'typical' human face. If a particular face has certain features that are stronger or weaker, bigger or smaller, longer or shorter, darker or fairer, than our 'typical' face, these are the items that we remember. In drawing a successful caricature, the artist has to know intuitively which features we have selected in this way, and he then has to super-normalize the strong points and sub-normalize the weak ones. The process is fundamentally the same as that employed in the drawings of children and primitive peoples, except that the caricaturist is concerned primarily with individual differences.

The visual arts, throughout much of their history, have been pervaded by this device of stimulus extremism. Super-normal and sub-normal modifications abound in almost all the earlier art forms. As the centuries passed, however, realism came more and more to dominate European art. The painter and the sculptor became burdened with the task of recording the external world as precisely as possible. It was not until the last century, when science took over this formidable duty (with the development of photography), that artists were able to return to a freer manipulation of their subject-matter. They were slow to react at first, and although the chains were broken in the nineteenth century, it was not until the twentieth century that they were fully shaken off. During the past sixty years wave after wave of rebellion has occurred as stimulus extremism has reasserted itself more and more powerfully. The rule once again has become: magnify selected elements and eliminate others.

When paintings of the human face began to be manipulated in this way by modern artists, there was an outcry. The pictures were scorned as decadent lunacies, as if they reflected some new disease of twentieth-century life, instead of a return to art's more basic business of pursuing the Stimulus Struggle. The melodramatic exaggerations of human behaviour in theatrical productions, ballets and operas, and the extreme magnifications of human emotions expressed in songs and poems, were happily accepted, but it took some time to adjust to similar stimulus extremisms in the visual arts. When totally abstract paintings began to appear they were attacked as meaningless by people who were perfectly willing to enjoy the total abstraction of any musical performance. But music had never been forced into the aesthetic strait-jacket of portraying natural sounds.

I have defined a super-normal stimulus as an artificial exaggeration of a natural stimulus, but the concept can also be applied in a special way to an invented stimulus. Let me take two clear-cut cases. The pink lips of a beautiful girl are, without any question, a perfectly natural, biological stimulus. If she exaggerates them by painting them a brighter pink, she is obviously converting them into a super-normal stimulus. There the issue is simple, and it is this sort of example I have been concentrating on up to now. But what about the sight of a shiny new motor car? This can be very stimulating, too, but it is an entirely artificial, invented stimulus. There is no natural, biological model against which we can compare it to find out if it has been super-normalized. And yet, as we look around at various motor cars, we can easily pick out some that seem to have the quality of being super-normal. They are bigger and more dramatic than most of the others. Manufacturers of motor cars are, in fact, just as concerned with producing super-normal stimuli as manufacturers of lipsticks. The situation is more fluid, because there is no natural, biological base-line against which to work; but the process is essentially the same. Once a new stimulus has been invented, it develops a base-line of its own. At any point in the history of motor cars it would be possible to produce a sketch of the typical, common and therefore 'normal' car of the period. It would also be possible to produce a sketch of the outstanding luxury motor car of the period which, at that time, was the super-normal vehicle. The only difference between this and the lipstick example is that the 'normal base-line' of the motor car changes with technical progress, whereas the natural pink lips stay the same.

The application of the super-normal principle is therefore widespread and penetrates almost all of our endeavours in one way or another. Freed from the demands of crude survival, we wring the last drop of stimulation out of anything we can lay our hands or eyes on. The result is that we sometimes get stimulus indigestion. The snag with making stimuli more powerful is that we run the risk of exhausting ourselves by the strength of our response. We become jaded. We begin to agree with the Shakespearean comment that

To gild refined gold, to paint the lily,
To throw a perfume on the violet ...
Is wasteful and ridiculous excess.

But at the same time we are forced to admit, with Wilde, that 'Nothing succeeds like excess.' So what do we do? The answer is that we bring into operation yet another subsidiary principle of the Stimulus Struggle:

This states that because super-normal stimuli are so powerful and our response to them can become exhausted, we must from time to time vary the elements that are selected for magnification. In other words, we ring the changes. When a switch of this sort occurs it is usually dramatic, because a whole trend is reversed. It does not, however, stop a particular branch of the Stimulus Struggle from being pursued, it merely shifts the points of super normal emphasis. Nowhere is this more clearly illustrated than in the world of fashionable clothing and body adornment.

In female costumes, where sexual display is paramount, this has given rise to what fashion experts refer to as the Law of Shifting Erogenous Zones. Technically, an erogenous zone is an area of the body that is particularly well supplied with nerve endings responsive to touch, direct stimulation of which is sexually arousing. The main areas are the genital region, the breasts, the mouth, the ear-lobes, the buttocks and the thighs. The neck, the arm-pits and the navel are sometimes added to the list. Female fashions are not, of course, concerned with tactile stimulation, but with the visual display (or concealment) of these sensitive areas. In extreme cases all these areas may be displayed at once, or, as in female Arab costume, all may be concealed. In the vast majority of super-tribal communities, however, some are displayed and others simultaneously concealed. Alternatively, some may be emphasized, although covered, while others are obliterated.

The Law of Shifting Erogenous Zones is concerned with the way in which concentration on one area gives way to concentration on another as time passes and fashions change. If the modern female emphasizes one zone for too long, the attraction wears off and a new super-normal shock is required to re-awaken interest.

In recent times the two main zones, the breasts and the pelvis, have remained largely concealed, but have been emphasized in various ways. One is by padding or tightening the clothing to exaggerate the shapes of these regions. The other is by approaching them as closely as possible with areas of exposed flesh. When this exposure creeps up on the breast region, with exceptionally low-cut costumes, it usually creeps away from the pelvic region, the dresses becoming longer. When the zone of interest shifts and the skirts become shorter, the neck-line rises. On occasions when bare midriffs have been popular, exposing the navel, the other zones have usually been rather well covered, often to the extent of the legs being concealed with some sort of trousers.

The great problem for fashion designers is that their super-normal stimuli are related to basic biological features. As there are only a few vital zones, this creates a strict limitation and forces the designers into a series of dangerously repetitive cycles. Only with great ingenuity can they overcome this difficulty. But there is always the head region to play with. Ear-lobes can be emphasized with ear-rings, necks with necklaces, the face with make-up. The Law of Shifting Erogenous Zones applies here too, and it is noticeable that when eye make-up becomes particularly striking and heavy, the lips usually become paler and less distinct.

Male fashion cycles follow a rather different course. The male in recent times has been more concerned with displaying his status than his sexual features. High status means access to leisure, and the most characteristic costumes of leisure are sporting clothes. Students of fashion history have unearthed the revealing fact that practically everything men wear today can be classified as 'ex-sports clothes'. Even our most formal attire can be shown to have these origins.

The system works like this. At any particular moment in recent history there has always been a highly functional costume to go with the high-status sport of the day. To wear such a costume indicates that you can afford the time and money to indulge in such a sport. This status display can be super-normalized by wearing the costume as ordinary day clothes, even when not pursuing the particular sport in question, thus magnifying the display by spreading it. The signals emanating from the sports clothes say, 'I am very leisured,' and they can say this almost as well for a non-sporting man who cannot afford to participate in the sport itself. After a while, when they have become completely accepted as everyday wear, they lose their impact. Then a new sport has to be raided for its unusual costume.

Back in the eighteenth century, English country gentlemen were exhibiting their status by taking to the hunting field. They adopted a sensible manner of dress for the occasion, wearing a coat that was cut away in the front, giving it the appearance of having tails at the back. They abandoned big floppy hats and began to wear stiff top hats, like prototype crash helmets. Once this costume was fully established as a high-status-sport outfit, it began to spread. At first it was the young bloods (the young swingers of the day) who started using a modified hunting costume as everyday wear. This was considered the height of daring, if not downright scandalous. But little by little the trend spread (young swingers get older), and by the middle of the nineteenth century the costume of top hat and tails had become normal everyday wear.

Having become so accepted and traditional, the top hat and tails had to be replaced with something new by the more daring members of society who wished to display their super-normal leisure signals. Other high-status sports available for raiding were shooting, fishing and golf. Billy-cock hats became bowlers and shooting tweeds became check lounge suits. The softer sporting hats became trilby hats. As the present century has advanced, the lounge suit has become more accepted as formal day wear and has become more sombre in the process. 'Morning dress', with top hat and tails, has been shifted one step further towards formality, being reserved now for special occasions such as weddings. It also survives as evening dress, but there the lounge suit has already caught up with it and stripped it of its tails to create a dinner-jacket suit.

Once the lounge suit had lost its daring, it had to be replaced, in its turn, by something more obviously sporting. Hunting may have dropped out of favour, but horse-riding in general still retained a high-status value, so here we go again. This time it was the hacking jacket that soon became known as a 'sports jacket'. Ironically it only acquired this name when it lost its true sporting function. It became the new casual wear for everyday use and still holds this position at the present time. Already, however, it is creeping into the more formal world of the business executive. Among the most daring dressers, it has even invaded that holy of holies, the formal evening occasion, in the guise of a patterned dinner jacket.

As the sports jacket spread into everyday life, the polo-necked sweater spread with it. Polo was another very high-status sport, and wearing the typical round-necked sweater of the game imparted instant status to the lucky wearer. But already this characteristic garment has lost its daring charm. A silk version of it was recently worn for the first time with a formal dinner jacket. Instantly shops were bombarded by young males clamouring for this latest sports attack on formality. It may have lost its impact as day wear, but as evening wear it was still able to shock, and its range spread accordingly.

Other similar trends have occurred during the last fifty years. Yachting blazers with brass buttons have been worn by men who have never stepped off dry land. Skiing suits have been worn by men (and women) who have never seen a snow-capped mountain. Just so long as a particular sport is exclusive and costly, it will be robbed for its costume signals. During the present century, leisure sports have been replaced to a certain extent by the habit of taking off for the sea-shores of warmer climates. This began with a craze for the French Riviera. Visitors there began copying the sweaters and shirts of the local fishermen. They were able to show that they had indulged in this expensive new status holiday by wearing modified versions of these shirts and sweaters back home. Immediately, a whole new range of casual clothes burst on to the market. In America, it became fashionable for wealthy, high-status males to own a ranch in the country, where they would dress in modified cowboy clothes. In no time at all, many a young ranchless city-dweller was striding along in his (further) modified cowboy suit. It could be argued that he took it straight from the Western movies, but this is unlikely. It would still have been fancy dress. However, once real, contemporary, high-status males are wearing it when they take their leisure, then all is well and a new take-over bid is on its way.

None of this, you may feel, explains the bizarre clothing of the way-out male teenager, who wears cravats, long hair, necklaces, coloured scarves, bracelets, buckled shoes, flared trousers and lace-cuffed shirts.

What kind of sport is *he* modifying? There is nothing

mysterious about the micro-skirted female teenager. All she has done, apart from shifting her erogenous zone to her thighs, is to take an emancipated leaf out of the male's fashion book, and steal a sports costume for everyday wear. The tennis skirt of the 1930s and the ice-skating skirt of the 1940s were already full-blooded micro-skirts. It only remained for some daring designer to modify them for everyday wear. But the flamboyant young male, what on earth is he doing? The answer seems to be that, with the recent setting up of a 'sub-culture of youth', it became necessary to develop an entirely new costume to go with it, one that owed as little as possible to the variations of the hated 'adult sub-culture'. Status in the 'youth sub-culture' has less to do with money and much more to do with sex appeal and virility. This has meant that the young males have begun to dress more like females, not because they are effeminate (a popular jibe of the older group), but because they are more concerned with sex attraction displays. In the recent past these have been largely the concern of the females, but now both sexes are involved. It is, in fact, a return to an earlier (pre-eighteenth century) condition of male dressing, and we should not be too surprised if the codpiece makes its reappearance any minute now. We may also see the return of elaborate male make-up. It is hard to say how long this phase will last because it will gradually be copied by older males who are already feeling disgruntled by the overt sex displays of their juniors. In returning to a peacock display, the young males of the 'youth sub-culture' have hit where it hurts most. The human male is in his sexually most potent condition at the age of sixteen to seventeen. By abandoning leisure status dress and replacing it with sex status dress, they have chosen the ideal weapon. However, as I said earlier, young bloods and young swingers grow older. It will be interesting to see what happens in twenty years' time, when there are bald Beatles in the board-room, and a new sub-culture of youth has arisen.

Almost everything we wear today, then, is the result of this Stimulus Struggle principle of ringing-the-changes to produce the shock effect of sudden novelty. What is daring today becomes ordinary tomorrow and stuffy the next day, and we quickly forget where it has all come from. How many men, climbing into their evening dress clothes and putting on their top hats, realize that they are donning the costume of a late eighteenth-century hunting squire? How many sombrely lounge-suited businessmen realize that they are following the dress of early nineteeenth-century country sportsmen? How many sports-jacketed young men think of themselves as horse-riders? How many open-neck-shirted, loose-knitted-sweatered young men think of themselves as Mediterranean fishermen? And how many mini-skirted young girls think of themselves as tennis-players or ice-skaters?

The shock is quickly over. The new style is quickly absorbed, and another one is then required to take its place and to provide a new stimulus. One thing we can always be sure of: whatever is today's most daring innovation in the world of fashion will become tomorrow's respectability, and will then rapidly fossilize into pompous formality as new rebellions crowd in to replace it. Only by this process of constant turn-over can the extremes of fashion, the super-normal stimuli of design, maintain their massive impact. Necessity may be the mother of invention, but where the super-normal stimuli of fashion are concerned it is also true to say that novelty is the mother of necessity.

Up to this point we have been considering the five principles of the Stimulus Struggle that are concerned with raising the behaviour output of the individual. Occasionally the reverse trend is called for. When this happens the sixth and final principle comes into operation:

6. If stimulation is too strong, you may reduce your behaviour output *by damping down responsiveness to incoming sensations.*

This is the cut-off principle. Some zoo animals find their confinement frightening and stressful, especially when they are newly arrived, moved to a fresh cage or housed with hostile or unsuitable companions. In their agitated condition they may suffer from abnormal over-stimulation. When this happens and they are unable to escape or hide, they must somehow switch off the incoming stimuli. They may do this simply by crouching in a corner and closing their eyes. This, at least, shuts off the visual stimuli. Excessive, prolonged sleeping (a device also used by invalids, both animal and human) also occurs as a more extreme form of cut-off. But they cannot crouch or sleep for ever.

While active, they can relieve their tensions to some extent by performing 'stereotypes'. These are small tics, repetitive patterns of twitching, rocking, jumping, swaying or turning, which, because they have become so familiar through being constantly repeated, have also become comforting. The point is that for the over-stimulated animal the environment is so strange and frightening that any action, no matter how meaningless, will have a calming effect so long as it is an old familiar pattern. It is like meeting an old friend in a crowd of strangers at a party. One can see these stereotypes going on all around the zoo. The huge elephants sway rhythmically back and forth; the young chimpanzee rocks its body to and fro; the squirrel leaps round and round in a tight circle like a wall-of-death rider; the tiger rubs its nose left and right across its bars until it is raw and bleeding.

If some of these over-stimulation patterns also occur from time to time in intensely bored animals, this is no accident, for the stress caused by gross under-stimulation is in some ways basically the same as the stress of over-stimulation. Both extremes are unpleasant and their unpleasantness causes a stereotyped response, as the animal tries desperately to escape back to the happy medium of moderate stimulation that is the goal of the Stimulus Struggle.

If the inmate of the human zoo becomes grossly over-stimulated, he too falls back on the cut-off principle. When many different stimuli are blaring away and conflicting with one another, the situation becomes unbearable. If we can run and hide, then all is well, but our complex commitments to super-tribal living usually prevent this. We can shut our eyes and cover our ears, but something more than blindfolds and ear-plugs are needed.

In extremis we resort to artificial aids. We take tranquillizers, sleeping pills (sometimes so many that we cut-off for good), over-doses of alcohol and a variety of drugs. This is a variant of the Stimulus Struggle which we can call Chemical Dreaming. To understand why, it will help to take a closer look at natural dreaming.

The great value of the process of normal, night-time dreaming is that it enables us to sort out and file away the chaos of the preceding day. Imagine an over-worked office, with mountains of documents, papers and notes pouring into it all through the day. The desks are piled high. The office workers cannot keep up with the incoming information and material. There is not enough time to file it away neatly before the end of the afternoon. They go home leaving the office in chaos. Next morning there will be another great influx and the situation will rapidly get out of hand.

If we are over-stimulated during the day, our brains taking in a mass of new information, much of it conflicting and difficult to classify, we go to bed in much the same condition as the chaotic office was left in at the end of the working day. But we are luckier than the over-worked office staff. At night-time someone comes into the office inside our skull and sorts everything out, files it neatly away and cleans up the office ready for the onslaught of the next day. In the brain of the human animal this process is what we call dreaming. We may obtain physical rest from sleep, but little more than we could get from lying awake all night. But awake we could not dream properly. The primary function of sleeping, then, is dreaming rather than resting our weary limbs. We sleep and we dream most of the night. The new information is sorted and filed and we awaken with a refreshed brain, ready to start the next day.

If day-time living becomes too frenzied, if we are too intensely over-stimulated, the ordinary dreaming mechanism becomes too severely tested. This leads to a pre-occupation with narcotics and the dangerous pursuit of Chemical Dreaming. In the stupors and trances of chemically induced states, we vaguely hope that the drugs will create a mimic of the dream-like state. But although they may be effective in helping to switch off the chaotic input from the outside world, they do not usually seem to assist in the positive dream function of sorting and filing. When they wear off, the temporary negative relief vanishes and the positive problem remains as it was before. The device is therefore doomed to be a disappointing one, with the addition of the possible anti-bonus of chemical addiction.

Another variation is the pursuit of what we can call Meditation Dreaming, in which the dream-like state is achieved by certain thought disciplines, yogal or otherwise. The cut-off, trance-like conditions produced by yoga, voodoo, hypnotism and certain magical and religious practices all have certain features in common. They usually involve sustained rhythmic repetition, either verbal or physical, and are followed by a condition of detachment from normal outside stimulation. In this way they can help to cut down the massive and usually conflicting input that is being suffered by the over-stimulated individual. They are therefore similar to the various forms of Chemical Dreaming, but as yet we have little information about the way they may, in addition, provide positive benefits of the kind we all enjoy when dreaming normally.

If the human fails to escape from a prolonged state of over-stimulation, he is liable to fall sick, mentally or physically. Stress diseases or nervous breakdowns may, for the luckier ones, provide their own cure. The invalid is forced, by his incapacity, to switch off the massive input. His sick-bed becomes his animal hiding-place.

Individuals who know they are particularly prone to over-stimulation often develop an early-warning signal. An old injury may start to play up, tonsils may swell, a bad tooth may throb, a skin rash may break out, a small twitch may reappear or a headache may begin again. Many people have a minor weakness of this kind which is really more of an old friend than an old enemy, because it warns them that they are 'over-doing' things and had better slow down if they wish to avoid something worse. If, as often happens, they are persuaded to have their particular weakness 'cured', they need have little fear of losing the early-warning advantage it bestowed on them; some other symptom will in all probability soon emerge to take its place. In the medical world this is sometimes known as the 'shifting syndrome'.

It is easy enough to understand how the modern super-tribesmen can come to suffer from this over-burdened state. As a species, we originally became intensely active and exploratory in connection with our special survival demands. The difficult role our hunting ancestors had to play insisted on it. Now, with the environment extensively under control, we are still saddled with our ancient system of high activity and high curiosity. Although we have reached a stage where we could easily afford to lie back and rest more often and more lengthily, we simply cannot do it. Instead we are forced to pursue the Stimulus Struggle. Since this is a new pursuit for us, we are not yet very expert performers and we are constantly going either too far or not far enough. Then, as soon as we feel ourselves becoming over-stimulated and over-active or under-stimulated and under-active, we veer away from the one painful extreme or the other and indulge in actions that tend to bring us back to the happy medium of optimum stimulation and optimum activity. The successful ones hold a steady centre course; the rest of us swing back and forth on either side of it.

We are helped to a certain extent by a slow process of adjustment. The countryman, living a quiet and peaceful life, develops a tolerance to this low level of activity. If a busy townsman were suddenly thrust into all that peace and quiet, he would quickly find it unbearably boring. If the countryman were thrown into the hurly-burly of chaotic town life, he would soon find it painfully stressful. It is fine to have a quiet week-end in the country as a de-stimulator, if you are a townsman, and it is great to have a day up in town as a stimulator, if you are a countryman. This obeys the balancing principles of the Stimulus Struggle; but much longer, and the balance is lost.

It is interesting that we are much less sympathetic towards a man who fails to adjust to a low level of activity than we are to one who fails to adjust to a high level. A bored and listless man annoys us more than a harassed and overburdened one. Both are failing to tackle the Stimulus Struggle efficiently. Both are liable to become irritable and bad-tempered, but we are much more prone to forgive the over-worked man. The reason for this is that pushing the level up a little too high is one of the things that keeps our cultures advancing. It is the intensely over-exploratory individuals who will become the great innovators and will change the face of the world in which we live. Those who pursue the Stimulus Struggle in a more balanced and successful way will also, of course, be exploratory, but they will tend to provide new variations on old themes rather than entirely new themes. They will also be happier, better adjusted individuals.

You may remember that at the outset I said the stakes of the game are high. What we stand to win or lose is our happiness, in extreme cases our sanity. The over-exploratory innovators should, according to this, therefore be comparatively unhappy and even show a tendency to suffer from mental illness. Bearing in mind the goal of the Stimulus Struggle, we should predict that, despite their greater achievements, such men and women must frequently live uneasy and discontented lives. History tends to confirm that this is so. Our debt to them is sometimes paid in the form of the special tolerance we show towards their frequently moody and wayward behaviour. We intuitively recognize that it is an inevitable outcome of the unbalanced way in which they are pursuing the Stimulus Struggle. As we shall see in the next chapter, however, we are not always so understanding.

CHAPTER SEVEN

THE CHILDLIKE ADULT

In many respects the play of children is similar to the Stimulus Struggle of adults. The child's parents take care of its survival problems and it is left with a great deal of surplus energy. Its playful activities help to burn up this energy. There is, however, a difference. We have seen that there are various ways of pursuing the adult Stimulus Struggle, one of which is the invention of new patterns of behaviour. In play, this element is much stronger. To the growing child, virtually every action it performs is a new invention. Its naïvety in the face of the environment more or less forces it to indulge in a non-stop process of innovation. Everything is novel. Each bout of playing is a voyage of discovery: discovery of itself, its abilities and capacities, and of the world about it. The development of inventiveness may not be the specific goal of play, but it is nevertheless its predominant feature and its most valuable bonus.

The explorations and inventions of childhood are usually trivial and ephemeral. In themselves they mean little. But if the processes they involve, the sense of wonder and curiosity, the urge to seek and find and test, can be prevented from fading with age, so that they remain to dominate the mature Stimulus Struggle, over-shadowing the less rewarding alternatives, then an important battle has been won: the battle for creativity.

Many people have puzzled over the secret of creativity. I contend that it is basically no more than the extension into adult life of these vital childlike qualities. The child asks new questions; the adult answers old ones; the childlike adult finds answers to new questions. The child is inventive; the adult is productive; the childlike adult is inventively productive. The child explores his environment; the adult organizes it; the childlike adult organizes his explorations and, by bringing order to them, strengthens them. He creates.

It is worth examining this phenomenon more closely. If a young chimpanzee, or a child, is placed in a room with a single familiar toy, he will play with it for a while and then lose interest. If he is offered, say, five familiar toys instead of only one, he will play first here, then there, moving from one to the other. By the time he gets back to the first one, the original toy will seem 'fresh' again and worthy of a little further attention. If, by contrast, an unfamiliar and novel toy is offered, it will immediately command his interest and produce a powerful reaction.

This 'new toy' response is the first essential of creativity, but it is one phase of the process. The strong exploratory urge of our species drives us on to investigate the new toy and to test it out in as many ways as we can devise. Once we have finished our explorations, then the unfamiliar toy will have become familiar. At this point it is our inventiveness that will come into action to utilize the new toy, or what we have learned from it, to set up and solve new problems. If, by re-combining our experiences from our different toys, we can make more out of them than we started with, then we have been creative.

If a young chimpanzee is put in a room with an ordinary chair, for example, it starts out by investigating the object, tapping it, hitting it, biting it, sniffing it and clambering over it. After a while these rather random activities give way to a more structured pattern of activity. It may, for instance, start jumping over the chair, using it as a piece of gymnastic equipment. It has 'invented' a vaulting box, and 'created' a new gymnastic activity. It had learned to jump over things before, but not in quite this way. By combining its past experiences with the investigation of this new toy, it creates the new action of rhythmic vaulting. If, later on, it is offered more complex apparatus, it will build on these earlier experiences again, incorporating the new elements.

This developmental process sounds very simple and straightforward, but it does not always fulfil its early promise. As children we all go through these processes of exploration, invention and creation, but the ultimate level of creativity we rise to as adults varies dramatically from individual to individual. At the worst, if the demands of the environment are too pressing, we stick to limited activities we know well. We do not risk new experiments. There is no time or energy to spare. If the environment seems too threatening, we would rather be sure than sorry: we fall back on the security of tried and trusted, familiar routines. The environmental situation has to change in one way or another before we will risk becoming more exploratory. Exploration involves uncertainty and uncertainty is frightening. Only two things will help us to overcome these fears. They are opposites: one is disaster, and the other is greatly increased security. A female rat, for instance, with a large litter to rear, is under heavy pressure. She works non-stop to keep her offspring fed, cleaned and protected. She will have little time for exploring. If disaster strikes—if her nest is flooded or destroyed—she will be forced into panic exploration. If, on the other hand, her young have been successfully reared and she has built up a large store of food, the pressure relaxes and, from a position of greater security, she is able to devote more time and energy to exploring her environment.

There are, then, two basic kinds of exploration: panic

exploration and security exploration. It is the same for the human animal. During the chaos and upheaval of war, a human community may be driven to inventiveness to surmount the disasters it faces. Alternatively, a successful, thriving community may be highly exploratory, striking out from its strong position of increased security. It is the community that is just managing to scrape along that will show little or no urge to explore.

Looking back on the history of our species it is easy to see how these two types of exploration have helped human progress to stumble on its way. When our early ancestors left the comforts of a fruit-picking, forest existence and took to open country, they were in serious difficulties. The extreme demands of the new environment forced them to be exploratory or die. Only when they had evolved into efficient, co-operative hunters did the pressure ease a little. They were at the 'scraping by' stage again. The result was that this condition lasted for a very long time, thousands upon thousands of years, with advances in technology occurring at an incredibly slow rate, simple developments in such things as implements and weapons, for example, taking hundreds of years to take a small new step.

Eventually, when primitive agriculture slowly emerged and the environment came more under our ancestors' control, the situation improved. When this was particularly successful, urbanization developed and a threshold was passed into a realm of new and dramatically increased social security. With it came a rush of the other kind of exploration—security exploration. This, in its turn, led to more and more startling developments, to more security and more exploration.

Unfortunately this was not the whole story. Man's rise to civilization would be a much happier tale if only it had been. But, sadly, events moved too fast and, as we have seen throughout this book, the pendulum of success and disaster began to swing crazily back and forth. Because we unleashed so much more than we were biologically equipped to cope with, our magnificent new social developments and complexities were as often abused as they were used. Our inability to deal rationally with the super-status and super-power that our super-tribal condition thrust upon us, led to new, more sudden and more challenging disasters than we had ever known. No sooner had a super-tribe settled down to a phase of great prosperity, with security exploration operating at full intensity, and wonderful new forms of creativity blossoming out, than something went wrong. Invaders, tyrants and aggressors smashed the delicate machinery of the intricate new social structures, and panic exploration was back on a major scale. For each new invention of construction, there was another of destruction, back and forth, back and forth, for ten thousand years, and it still goes on today. It is the horror of atomic weapons that has given us the glory of atomic energy, and it is the glory of biological research that may yet give us the horror of biological warfare.

In between these two extremes there are still millions of people living the simple lives of primitive agriculturalists, tilling the soil much as our early ancestors did. In a few areas primitive hunters survive. Because they have stayed at the 'scraping by' stage, they are typically non-exploratory. Like the left-over great apes—the chimpanzees, the gorillas and the orang-utans—they have the potential for inventiveness and exploration, but it is not called forth to any appreciable extent. Experiments with chimpanzees in captivity have revealed how quickly they can be encouraged to develop their exploratory potential: they can operate machines, paint pictures and solve all kinds of experimental puzzles; but in the wild state they do not even learn to build crude shelters to keep out the rain. For them, and the simpler human communities, the scraping-by existence—not too difficult and not too easy—has blunted their exploratory urges. For the rest of us, one extreme follows the other and we constantly explore from either an excess of panic or an excess of security.

From time to time there are those among us who cast an envious backward glance at the 'simple life' of primitive communities and start to wish we had never left our primeval Forest of Eden. In some cases, serious attempts have been made to convert such thoughts into actions. Much as we may sympathize with these projects, it must be realized that they are fraught with difficulties. The inherent artificiality of pseudo-primitive drop-out communities, such as those that have appeared in North America and elsewhere recently, is a primary weakness. They are, after all, composed of individuals who have tasted the excitements of super-tribal life as well as its horrors. They have been conditioned throughout their lives to a high level of mental activity. In a sense they have lost their social innocence, and the loss of innocence is an irreversible process.

At first, all may go well for the neo-primitive, but this is deceptive. What happens is that the initial return to the simple way of life throws up an enormous challenge to the ex-inmate of the human zoo. His new role may be simple in theory, but in practice it is full of fascinatingly novel problems. The establishment of a pseudo-primitive community by a group of ex-city-dwellers becomes, in fact, a major exploratory act. This, rather than the official return to pure simplicity, is what makes the project so satisfying, as any Boy Scout will testify. But what happens once the initial challenge has been met and overcome? Whether it is a remote, rural or cave-dwelling group, or whether it is a self-insulated, pseudo-primitive group set up in an isolated pocket inside the city itself, the answer is the same. Disillusionment sets in, as the monotony begins to assail the brains that have been irreversibly trained to the higher, super-tribal level. Either the group collapses, or it starts to stir itself into action. If the new activity is successful, then the community will soon find itself becoming organized and expanding. In no time at all it will be back in the super-tribal rat-race.

It is difficult enough in the twentieth century to remain as a genuine primitive community, like the Eskimoes or aborigines, let alone a pseudo-primitive one. Even the traditionally resistant European gypsies are gradually succumbing to the relentless spread of the human zoo condition.

The tragedy for those who wish to solve their problems by a return to the simple life is that, even if they somehow contrived to 'de-train' their highly activated brains, such individuals would still remain extremely vulnerable in their small rebel communities. The human zoo would find it hard to leave them alone. They would either be exploited as a tourist attraction, as so many of the genuine primitives are today, or, if they become an irritant, they would be attacked and disbanded. There is no escape from the super-tribal monster and we may as well make the best of it.

If we are condemned to a complex social existence, as it seems we are, then the trick is to ensure that *we* make use of *it,* rather than let *it* make use of *us*. If we are going to be forced to pursue the Stimulus Struggle, then the important thing is to select the most rewarding method of approach. As I have already indicated, the best way to do this is to give priority to the inventive, exploratory principle, not inadvertently like the drop-outs, who find themselves all too soon in an exploratory blind alley, but deliberately, gearing our inventiveness to the mainstream of our super-tribal existence.

Given the fact that each super-tribesman is free to choose which way he pursues the Stimulus Struggle, it remains to ask why he does not select the inventive solution more frequently. With the enormous exploratory potential of his brain lying idle and with his experience of inventive playfulness in childhood behind him, he should in theory favour this solution above all others. In any thriving super-tribal city *all* the citizens should be potential 'inventors'. Why, then, do so few of them indulge in active creativity, while the others are satisfied to enjoy their inventions second-hand, watching them on television, or are content to play simple games and sports with strictly limited possibilities for inventiveness? They all appear to have the necessary background for becoming childlike adults. The super-tribe, like a gigantic parent, protects them and cares for them, so why is it that they do not all develop bigger and better childlike curiosity?

Part of the answer is that children are subordinate to adults. Inevitably, dominant animals try to control the behaviour of their subordinates. Much as adults may love their children, they cannot help seeing them as a growing threat to their dominance. They know that with ultimate senility they will have to give way to them, but they do everything they can to postpone the evil day. There is therefore a strong tendency to suppress inventiveness in members of the community younger than oneself. An appreciation of the value of their 'fresh eyes' and their new creativeness works against this, but it is an uphill struggle. By the time the new generation has matured to the point where its members could be wildly inventive, childlike adults, they are already burdened with a heavy sense of conformity. Struggling against this as hard as they can, they in turn are then faced with the threat of another younger generation coming up beneath them, and the suppressive process repeats itself. Only those rare individuals who experience an unusual childhood, from this point of view, will be able to achieve a level of great creativity in adult life. How unusual does such a childhood have to be? It either has to be so suppressive that the growing child revolts against the traditions of its elders in a big way (many of our greatest creative talents were so-called delinquent children), or it has to be so un-suppressive that the heavy hand of conformity rests only lightly on its shoulder. If a child is strongly punished for its inventiveness (which, after all, is essentially rebellious in nature), it may spend the rest of its adult live making up for lost time. If a child is strongly rewarded for its inventiveness, then it may never lose it, no matter what pressures are brought to bear on it in later years. Both types can make a great impact in adult society, but the second will probably suffer less from obsessive limitations in his creative acts.

The vast majority of children will, of course, receive a more balanced mixture of punishment and reward for their inventiveness and will emerge into adult life with personalities that are both moderately creative and moderately conformist. They will become the adult-adults. They will tend to read the newspapers rather than make the news that goes into them. Their attitude to the childlike adults will be ambivalent; on the one hand they will applaud them for providing the much-needed sources of novelty, but on the other they will envy them. The creative talent will therefore find himself alternately praised and damned by society in a bewildering way, and will be constantly in doubt about his acceptance by the rest of the community.

Modern education has made great strides in encouraging inventiveness, but it still has a long way to go before it can completely rid itself of the urge to suppress creativity. It is inevitable that bright young students will be seen as a threat by older academics, and it requires great self-control for teachers to overcome this. The system is designed to make it easy, but their nature, as dominant males, is not. Under the circumstances it is remarkable that they manage to control themselves as well as they do. There is a difference here between the school level and the university level. In most schools the dominance of the master over his pupils is strongly and directly expressed, both socially and intellectually. He uses his greater experience to conquer their greater inventiveness. His brain has probably become more rigidified than theirs, but he masks this weakness by imparting large quantities of 'hard' facts. There is no argument, only instruction. (The situation is improving and there are, of course, exceptions, but this still applies as a general rule.)

At the university level the scene changes. There are many more facts to be handed down, but they are not quite so 'hard'. The student is now expected to question them and assess them, and eventually to invent new ideas of his own. But at both stages, the school and the university, there is something else going on beneath the surface, something that has little to do with the encouragement of intellectual expansion, but a geat deal to do with the indoctrination of super-tribal identity. To understand this we must look at what happened in simpler tribal societies.

In many cultures children at puberty have been subjected to impressive initiation ceremonies. They are taken away from their parents and kept in groups. They are then forced to undergo severe ordeals, often involving torture of mutilation. Operations are performed on their genitals, or their bodies may be scarred, burned, whipped or stung with ants. At the same time they are instructed in the secrets of the tribe. When the rituals are over, they are accepted as adult members of the society.

Before we see how this relates to the rituals of modern education, it is important to ask what value these seemingly damaging activities have. In the first place they isolate the maturing child from its parents. Previously it could always run to them for comfort when in pain. Now, for the first time, the child must endure pain and fear in a situation in which its parents cannot be called upon for assistance. (The initiation ceremonies are usually performed in strict privacy by the tribal elders and the rest of the tribe is excluded.) This helps to smash the child's sense of dependence on its parents and shift its allegiance away from the family home and on to the tribal community as a whole. The fact that at the same time he is allowed to share the adult tribal secrets strengthens the process by giving a fabric to his new tribal identity. Secondly, the violence of the emotional experience accompanying his instruction helps to burn the details of the tribal teachings into his brain. Just as we find it impossible to forget the details of a traumatic experience, such as a motor accident, so the tribal initiate will remember to his dying day the secrets that were imparted to him on that frightening occasion. The initiation is, in a sense, contrived traumatic teaching. Thirdly, it makes it absolutely clear to the sub-adult that, although he is now joining the ranks of his elders, he is doing so very much in the role of a subordinate. The intense power which they exercised over him will also be vividly remembered.

Modern schools and universities may not sting their students with ants but, in many ways, the educational system today shows striking similarities to the earlier tribal initiation procedures. To start with, the children are taken away from their parents and put in the hands of super-tribal elders—the academics—who instruct them in the 'secrets' of the super-tribe. In many cultures they are still made to wear a separate uniform, to set them apart and strengthen their new allegiance. They may also be encouraged to indulge in certain rituals, such as singing school or college songs. The severe ordeals of the tribal initiation ceremony no longer leave physical scars. (German duelling scars never really caught on.) But physical ordeals of a less vicious sort have persisted almost everywhere until very recently, at least at the school level, in the form of buttock-caning. Like the genital mutilations of the tribal ceremonies, this form of punishment has always had a sexual flavour and cannot be dissociated from the phenomenon of Status Sex.

In the absence of a more violent form of ordeal stemming from the teachers, the older pupils frequently take over the role of 'tribal elders' and administer their own tortures on 'new boys'. These vary from place to place. At one school, for instance, newcomers are 'grassed', having bundles of grasses stuffed inside their clothing. At another they are 'stoned', being bent over a large stone and spanked. At another they are forced to run down a long corridor lined with older pupils who kick them as they pass. At yet another they are 'bumped', being held by the arms and legs and banged on the ground as many times as their age in years. Alternatively on the day a new pupil wears his first school uniform, he may have his flesh pinched once for each new article of clothing, by each senior pupil. In rare cases, the ordeal is much more elaborate and may almost approach a full-scale tribal initiation ceremony. Even today, occasional deaths are reported as a result of these activities.

Unlike the primitive tribal situation, there is nothing to stop a tortured boy from complaining to his parents, but this hardly ever happens because it would bring shame to the boy in question. Many parents are not even aware of the trials their children are undergoing. The ancient practice of alienating a child from its family home has already begun to work its strange magic.

Although these unofficial initiation rites have persisted here and there, the official punishment of caning by teachers has recently lost ground, due to the pressure of public opinion and the revised ideas of certain teachers. But if official ordeal by physical means is disappearing, there always remains the alternative of the mental ordeal. Throughout virtually the whole of the modern educational system there now exists one powerful and impressive form of super-tribal initiation ceremony, which goes under the revealing name of 'examinations'. These are conducted in the heavy atmosphere of high ritual, with the pupils cut off from all outside assistance. Just as in the tribal ritual, no one can help them. They must suffer on their own. At all other times in their lives they can make use of books of reference, or discussions over difficult points, when they are applying their intelligence to a problem, but not during the private rituals of the dreaded examinations.

The ordeal is further intensified by setting a strict time limit and by crowding all the different examinations together in the short space of a few days or weeks. The overall effect of these measures is to create a considerable amount of mental torment, again recalling the mood of the more primitive initiation ceremonies of simple tribes.

When the final exams are over, at university level, the students who have 'passed the test' become qualified as special members of the adult section of the super-tribe. They don elaborate display robes and take part in a further ritual called the degree ceremony, in the presence of the academic elders wearing their even more impressive and dramatic robes.

The university student phase usually lasts for three years, which is a long time, as initiation ceremonies go. For some, it is too long. The isolation from parental assistance and the comforting social environment of the home, coupled with the looming demands of the examination ordeal, often proves too stressful for the young initiate. At British universities roughly twenty per cent of the undergraduates seek psychiatric assistance at some time during their three-year course of study. For some, the situation becomes unbearable and suicides are unusually frequent, the university rate being three to six times higher than the national average for the same age group. At Oxford and Cambridge universities, the suicide rate is seven to ten times higher.

Clearly the educational ordeals I have been describing have little to do with the business of encouraging and expanding childhood playfulness, inventiveness and creativity. Like the primitive tribal initiation ceremonies, they have instead to do with the indoctrination of super-tribal identity. As such they play an important cohesive role, but the development of the creative intellect is another matter altogether.

One of the excuses given for the ritual ordeals of modern education is that they provide the only way of ensuring that the students will absorb the huge mass of facts now available. It is true that detailed knowledge and specialist skills are necessary today, before an adult can even begin to be confidently inventive. Also, the examination ceremonies prevent cheating. Furthermore, it could be claimed that students should deliberately be subjected to stress to test their stamina. The challenges of adult life are also stressful, and if a student cracks up under the strain of educational ordeals, then he probably was not equipped to withstand the post-educational pressures either. These arguments are plausible, and yet one still senses the crushing of creative potentials under the heavy boot of educational ritual procedures. It is undeniable that the present system is a considerable advance over earlier educational methods, and that for those who survive the ordeals there is a great deal of exploratory nourishment to be gained. Our super-tribes today contain more successful childlike adults than ever before. Yet despite this, in many spheres there still exists an oppressive atmosphere of emotional resistance to radically new, inventive ideas. Dominant individuals encourage minor inventiveness in the form of new variations on old themes, but resist major inventiveness that leads to entirely new themes.

To give an example: it is astonishing the way we go on and on trying to improve something as primitive as the engine used in present-day motor vehicles. There is a strong chance that by the twenty-first century it will have become as obsolete as the horse and cart is today. That this is only a strong chance and not a complete certainty is due entirely to the fact that at this moment all the best brains in the profession are busily absorbed by the minor inventive problems of how to achieve minute improvements in the performance of the existing machinery, rather than searching for something entirely new.

This tendency towards short-sightedness in adult exploratory behaviour is a measure of the insecurity of a peaceful society. Perhaps, as we move farther into the atomic age, we shall reach such peaks of super-tribal security, or hit such depths of super-tribal panic, that we shall become increasingly exploratory, inventive and creative.

It will not be an easy struggle, however, and recent events at universities all over the world bear this out. The improved educational systems have already been so effective that many students are no longer prepared to accept without question the authority of their elders. The community was not ready for this and has been taken by surprise. The result is that when groups of students indulge in noisy protest, society is outraged. The educational authorities are horrified. The ingratitude of it all! What has gone wrong?

If we are ruthlessly honest with ourselves, the answer is not hard to find. It is contained in the official doctrines of these same educational authorities. As they face the upheaval, they must contemplate the uncomfortable fact that they have brought it on themselves. They literally asked for it. 'Think for yourselves,' they said, 'be resourceful, be active, be inventive.' Contradicting themselves in the same breath, they added: 'But do it on *our* terms, in *our* way and above all abide by *our* rituals.'

It should be obvious, even to a senile authority, that the more the first message is obeyed, the more the second will be ignored. Unfortunately the human animal is remarkably good at blinding itself to the obvious if it happens to be particularly unappealing, and it is this self-blinding process that has caused so many of the present difficulties.

When they called for increased resourcefulness and inventiveness, the authorities did not anticipate the magnitude of the response that was to follow, and it quickly got out of control. They did not seem to realize that they were encouraging something which already had a strong biological backing. They mistakenly treated resourcefulness and a sense of creative responsibility as properties alien to the young brain, when in fact they were hidden there all the time, only waiting to burst out if they got the chance.

The old-fashioned educational systems, as I have already pointed out, had done their best to suppress these properties, demanding a much greater obedience to the established rules of the elders. They had rigorously imposed parrot-fashion learning of rigid dogmas. Inventiveness had been forced to fight its own battles, pushing its way to the surface only in exceptional, isolated individuals. When it did manage to break through, however, its value to society was indisputable, and this led eventually to the present-day movement on the part of the establishment actively to encourage it. Approaching the matter rationally, they saw inventiveness and creativity as immense aids to greater social progress. At the same time, the deeply ingrained urge of these super-tribal authorities to retain their vice-like grip on the social order still persisted, making them oppose the very trend which they were now officially supporting. They entrenched ever more firmly, moulding society into a shape that was guaranteed to resist the new waves of inventiveness that they themselves had unleashed. A collision was inevitable.

The initial response of the establishment, as the mood of experimentation grew, was one of tolerant amusement. Cautiously viewing the younger generation's increasingly daring assaults on the accepted traditions of the arts, literature, music, entertainment and social custom, they kept their distance. Their tolerance collapsed, however, as this trend spread into the more threatening areas of politics and international affairs.

As isolated, eccentric thinkers grew into a massive, querulous crowd, the establishment switched hurriedly to its most primitive form of response—attack. The young intellectual, instead of being patted tolerantly on the head, found himself struck on the skull by a police baton. The lively brains that society had so carefully nurtured were soon suffering not from strain but from concussion.

The moral for the authorities is clear: do not give creative liberties unless you expect people to take them. The young human animal is not a stupid, idle creature that has to be driven to creativity; it is a fundamentally creative being that in the past has been made to appear idle by the suppressive influences exerted on it from above. The establishment's reply is that dissenting students are bent, not on positive innovation, but on negative disruption. Against this, however, it can be argued that these two processes are very closely related and that the former only degenerates into the latter when it finds itself blocked.

The secret is to provide a social environment capable of absorbing as much inventiveness and novelty as it sets out to encourage in the first place. As the super-tribes are constantly swelling in size and the human zoo is being ever more cramped and crowded, this requires careful and imaginative planning. Above all, it calls for considerably more insight into the biological demands of the human species, on the part of politicians, administrators and city planners, than has been evident in the recent past.

The more closely one looks at the situation, the more alarming it becomes. Well-meaning reformers and organizers busily work towards what they consider to be improved living conditions, never for one moment doubting the validity of what they are doing. Who, after all, can deny the value of providing more houses, more flats, more cars, more hospitals, more schools and more food? If perhaps there is a degree of sameness about all these bright new commodities, this cannot be helped. The human population is growing so fast that there is not enough time or space to do it any better. The snag is that while, on the one hand, all those new schools are bursting with pupils, inventiveness at the ready, dead set to *change* things, the other new developments are conspiring to render startling new innovations more and more impossible. In their ever-expanding and highly regimented monotony, these developments unavoidably favour widespread indulgence in the more trivial solutions to the Stimulus Struggle. If we are not careful, the human zoo will become increasingly like a Victorian menagerie, with tiny cages full of twitching, pacing captives.

Some science-fiction writers take the pessimistic view.

When depicting the future, they portray it as an existence in which human individuals are subjected to a suffocating degree of increased uniformity, as if new developments have brought further invention almost to a standstill. Everyone wears drab tunics and automation dominates the environment. If new inventions do occur, they only serve to squeeze the trap tighter around the human brain.

It could be argued that this picture merely reflects the paucity of the writers' imaginations, but there is more to it than that. To some extent they are simply exaggerating the trend that can already be detected in present-day conditions. They are responding to the relentless growth of what has been called the 'planner's prison'. The trouble is that as new developments in medicine, hygiene, housing and food-production make it possible to cram more and more people efficiently into a given space, the creative elements in society become more and more side-tracked into problems of quantity rather than quality. Precedence is given to those inventions that permit further increases in repetitive mediocrity. Efficient homogeneity takes precedence over stimulating heterogeneity.

As one rebel planner pointed out, a straight path between two buildings may be the most efficient (and the cheapest) solution, but that does not mean that it is the *best* path from the point of view of satisfying human needs. The human animal requires a spatial territory in which to live that possesses unique features, surprises, visual oddities, landmarks and architectural idiosyncrasies. Without them it can have little meaning. A neatly symmetrical, geometric pattern may be useful for holding up a roof, or for facilitating the prefabrication of mass-produced housing-units, but when such patterning is applied at the landscape level, it is going against the nature of the human animal. Why else is it so much fun to wander down a twisting country lane? Why else do children prefer to play on rubbish dumps or in derelict buildings, rather than on their immaculate, sterile, geometrically arranged playgrounds?

The current architectural trend towards austere design-simplicity can easily get out of hand and be used as an excuse for lack of imagination. Minimal aesthetic statements are only exciting as contrasts to other, more complex statements. When they come to dominate the scene, the results can be extremely damaging. Modern architecture has been heading this way for some time, strongly encouraged by the human zoo planners. Huge tower blocks of repetitive, uniform apartments have proliferated in many cities as a response to the housing demands of the swelling super-tribal populations. The excuse has been slum clearance, but the results have all too often been the creation of the super-slums of the immediate future. In a sense they are worse than nothing because, since they falsely give the impression of progress, they create complacency and deaden the chance of a genuine advance.

The more enlightened animal zoos have been getting rid of their old monkey houses. The zoo directors saw what was happening to the inmates, and realized that putting more hygienic tiles on the walls and improving the drainage was no real solution. The directors of human zoos, faced with mushrooming populations, have not been so far-sighted. The outcome of their experiments in high-density uniformity are now being assessed in the juvenile courts and psychiatrists' consulting rooms. On some housing estates it has even been recommended that prospective tenants of upper-storey apartments should undergo psychiatric examination *before* taking up residence, to ascertain whether, in the psychiatrist's opinion, they will be able to stand the strain of their brave new way of live.

This fact alone should provide sufficient warning to the

planners, revealing to them clearly the enormity of the folly they are committing, but as yet there is little sign that they are heeding such warnings. When confronted with the shortcomings of their endeavours, they reply that they have no alternative; there are more and more people and they have to be housed. But somehow alternatives must be found. The whole nature of city-complexes must be re-examined. The harassed citizens of the human zoo must in some way be given back the 'village-community' feeling of social identity. A genuine village, seen from the air, looks like an organic growth, not a piece of slide-rule geometry, a point which most planners seem studiously to ignore. They fail to appreciate the basic demands of human territorial behaviour. Houses and streets are not primarily for looking at, like set pieces, but for moving about in. The architectural environment should make its impact second by second and minute by minute as we travel along our territorial tracks, the pattern changing subtly with each new line of vision. As we turn a corner, or open a door, the last thing our navigational sense wants to be faced with is a spatial configuration that drearily duplicates the one we have just left. All too frequently, however, this is precisely what happens, the architectural planner having loomed over his drawing-board like a bomber-pilot sighting a target, rather than attempting to project himself down as a small mobile object travelling around *inside* the environment.

These problems of repetitive monotony and uniformity do, of course, permeate almost all aspects of modern living. With the ever increasing complexity of the human zoo environment, the dangers of intensified social regimentation are mounting daily. While organizers struggle to encase human behaviour in a more and more rigid framework, other trends work in the opposite direction. As we have seen, the steadily improving education of the young and the growing affluence of their elders both lead to a demand for more and more stimulation, adventure, excitement and experimentation. If the modern world fails to permit these trends, then tomorrow's super-tribesmen will fight hard to change that world. They will have the training and the time and the exploratory energy to do so, and somehow they will manage it. If they feel themselves trapped in a planner's prison they will stage a prison riot. If the environment does not permit creative innovations, they will smash it in order to be able to start again. This is one of the greatest dilemmas our societies face. To resolve it is our formidable task for the future.

Unfortunately we tend to forget that we are animals with certain specific weaknesses and certain specific strengths. We think of ourselves as blank sheets on which anything can be written. We are not. We come into the world with a set of basic instructions and we ignore or disobey them at our peril.

The politicians, the administrators and the other super-tribal leaders are good social mathematicians, but this is not enough. In what promises to be the ever more crowded world of the future, they must become good biologists as well, because somewhere in all that mass of wires, cables, plastics, concrete, bricks, metal and glass which they control, there is an animal, a human animal, a primitive tribal hunter, masquerading as a civilized, super-tribal citizen and desperately struggling to match his ancient inherited qualities with his extraordinary new situation. If he is given the chance he may yet contrive to turn his human zoo into a magnificent human game-park. If he is not, it may proliferate into a gigantic lunatic asylum, like one of the hideously cramped animal menageries of the last century.

For us, the super-tribesmen of the twentieth century, it will be interesting to see what happens. For our children, however, it will be more then merely interesting. By the time they are in charge of the situation, the human species will no doubt be facing problems of such magnitude that it will be a matter of living or dying.

APPENDIX

LITERATURE

It is impossible to list all the many works that have been of assistance in writing *The Human Zoo*. I have therefore included only those which have either been important in providing information on a specific point, or are of particular interest for further reading. They are arranged below on a chapter-by-chapter and topic-by-topic basis. From the names and dates given, it is possible to trace the full references in the bibliography that follows.

CHAPTER ONE. TRIBES AND SUPER-TRIBES

Home range of prehistoric man: Washburn and Devore, 1962.
Prehistoric man: Boule and Vallois, 1957. Clark and Piggott, 1965. Read, 1925. Tax, 1960. Washburn, 1962.
Farming origins: Cole, 1959. Piggott, 1965. Zeuner, 1963.
Urban origins: Piggott, 1961, 1965. Smailes, 1953.
Mourning dress: Crawley, 1931.

CHAPTER TWO. STATUS AND SUPER-STATUS

Behaviour of baboons: Hall and Devore, 1965.
Dominance patterns: Caine, 1960.
Status seekers: Packard, 1960.
Mimicry: Wickler, 1968.
Suicide: Berelson and Steiner, 1964. Stengel, 1964. Woddis, 1957.
Re-direction of aggression: Bastock, Morris and Moynihan, 1953.
Cruelty to animals: Jennison, 1937. Turner, 1964.

CHAPTER THREE. SEX AND SUPER-SEX

Sexual behaviour: Beach, 1965. Ford and Beach, 1952. Hediger, 1965. Kinsey *et al.*, 1948, 1953. Morris, 1956, 1964, 1966 and 1967.
Masturbation: Kinsey *et al.*, 1948.

Religious ecstacy: Bataille, 1962.
Boredom: Berlyne, 1960.
Displacement activities: Tinbergen, 1951.
Monkey prostitutes: Zuckerman, 1932.
Feline display: Leyhausen, 1956.
Sexual mimicry: Wickler, 1967.
Status sex: Russell and Russell, 1961.
Phallic symbols: Knight and Wright, 1957. Boullet, 1961.
Maltese cross: Adams, 1870.

CHAPTER FOUR. IN-GROUPS AND OUT-GROUPS

Aggression and War: Ardrey, 1963, 1967. Berkowitz, 1962. Carthy and Ebling, 1964. Lorenz, 1963. Richardson, 1960. Storr, 1968.
Races of man: Broca, 1864. Coon, 1963, 1966. Montagu, 1945. Pickering, 1850. Smith, 1968.
Racial conflict: Berelson and Steiner, 1964. Segal, 1966.
Population levels: Fremlin, 1965.

CHAPTER FIVE. IMPRINTING AND MAL-IMPRINTING

Imprinting in animals: Lorenz, 1935. Sluckin, 1965.
Mal-imprinting in animals: Hediger, 1950, 1965 (zoo animals). Morris, 1964 (zoo animals). Scott, 1956, 1958 (dogs). Scott and Fuller, 1965 (dogs). Whitman, 1919 (pigeons).
Social isolation in monkeys: Harlow and Harlow, 1962.
Human infant bonding: Ambrose, 1960. Brackbill and Thompson, 1967.
Pair-bonding: Morris, 1967.
Fetishism: Freeman, 1967. Hartwich, 1959.
Homosexuality: Morris, 1952, 1954, 1955. Schutz, 1965. West, 1968.
Pet-keeping: Morris and Morris, 1966.

CHAPTER SIX. THE STIMULUS STRUGGLE

Zoo animals: Appelman, 1960. Hediger, 1950. Inhelder, 1962. Lang, 1943. Lyall-Watson, 1963. Morris, 1962, 1964, 1966.
Boredom and stress: Berlyne, 1960.
Aesthetics: Morris, 1962.
Bestiality: Kinsey *et al.*, 1948, 1953.
Super-normal stimuli: Morris, 1956. Tinbergen, 1951, 1953.
Children's drawings: Morris, 1962.
Costume: Laver, 1950, 1952, 1963.
Cut-off: Chance, 1962.

CHAPTER SEVEN. THE CHILDLIKE ADULT

Chimpanzee curiosity: Morris, 1962. Morris and Morris, 1966.
Initiation ceremonies: Cohen, 1964.
School rituals: Opie and Opie, 1959.

BIBLIOGRAPHY

ADAMS, A. L., *Notes of a Naturalist in the Nile Valley and Malta* (Edmonston and Douglas, 1870)

AMBROSE, J. A., 'The smiling response in early human infancy' (Ph.D. thesis, London University, 1960), pp. 1–660

APPELMAN, F. J., 'Feeding of zoo animals by the public', in *Internat. Zoo Yearbook* 2 (1960), pp. 94–5

ARDREY, R., *African Genesis* (Atheneum, 1961)

ARDREY, R., *The Territorial Imperative* (Collins, 1967)

BASTOCK, M., D. MORRIS and M. MOYNIHAN, 'Some comments on conflict and thwarting in animals', in *Behaviour* 6 (1953), pp. 66–84

BATAILLE, G., *Eroticism* (Calder, 1962)

BEACH, F. A., *Sex and Behavior* (Wiley, 1965)

BERELSON, B., and G. A. STEINER, *Human Behavior* (Harcourt, Brace and World, 1964)

BERKOWITZ, L., *Aggression* (McGraw-Hill, 1962)

BERLYNE, D. E., *Conflict, Arousal and Curiosity* (McGraw-Hill, 1960)

BOULE, M., and H. V. VALLOIS, *Fossil Men* (Thames & Hudson, 1957)

BOULLET, J., *Symbolisme Sexuel* (Pauvert, 1961)

BRACKBILL, Y., and G. G. THOMPSON, *Behavior in Infancy and Early Childhood* (Free Press, 1967)

BROCA, P., *On the Phenomena of Hybridity in the Genus Homo* (Longman, Green, Longman & Roberts, 1864)

CAINE, M., *The S-Man* (Hutchinson, 1960)

CARTHY, J. D., and F. J. EBLING, *The Natural History of Aggression* (Academic Press, 1964)

CHANCE, M. R. A., 'An interpretation of some agonistic postures; the role of cut-off acts and postures', in *Symp. Zool. Soc. London* 8 (1962), pp. 71–89

CLARK, G., and S. PIGGOTT, *Prehistoric Societies* (Hutchinson, 1965)

COHEN, Y. A., *The Transition from Childhood to Adolescence* (Aldine, 1964)

COLE, S., *The Neolithic Revolution* (British Museum, 1959)

COON, C. S., *The Origin of Races* (Cape, 1963)

COON, C. S., *The Living Races of Man* (Cape, 1963)

CRAWLEY, E., *Dress, Drinks and Drums* (Methuen, 1931)

FORD, C. S., and F. A. BEACH, *Patterns of Sexual Behaviour* (Eyre & Spottiswoode, 1952)

FREEMAN, G., *The Undergrowth of Literature* (Nelson, 1967)

FREMLIN, J. H., 'How many people can the world support?' in *New Scientist* 24 (1965), pp. 285–7

HALL, K. R. L., and I. DEVORE, 'Baboon social behaviour', in *Primate Behavior* (Editor: I. Devore), (Holt, Rinehart, Winston, 1965)

HARLOW, H. H., and M. K. HARLOW, 'Social deprivation in monkeys', in *Sci. Amer.* 207 (1962) pp. 136–46

HARLOW, H. H., and M. K. HARLOW, 'The effect of rearing conditions on behaviour', in *Bull. Menninger Clin.* 26 (1962), pp. 213–24

HARTWICH, A., *Aberrations of Sexual Life* (After the *Psychopathia Sexualis* of Kraft-Ebing), (Staples Press, 1959)

HEDIGER, H., *Wild Animals in Captivity* (Butterworth, 1950)

HEDIGER, H., 'Environmental factors influencing the reproduction of zoo animals', in *Sex and Behaviour* (Editor: F. A. Beach), (Wiley, 1965)

INHELDER, E., 'Skizzen zu einer Verhaltenspathologie reactiver Störungen bei Tieren', in *Schweiz. Arch. Neurol. Psychiat.* 89 (1962), pp 276–326

JENNISON, G., *Animals for Show and Pleasure in Ancient Rome* (Manchester University Press, 1937)

KINSEY, A. C., W. B. POMEROY and C. E. MARTIN, *Sexual Behavior in the Human Male* (Saunders, 1948)

KINSEY, A. C., W. B. POMEROY, C. E. MARTIN and P. H. GEBHARD, *Sexual Behavior in the Human Female* (Saunders, 1953)

KNIGHT, R. P., and T. WRIGHT, *Sexual Symbolism* (Julian Press, 1957)

LANG, E. M., 'Eine ungewöhnliche Stereotype bei einem Lippenbären', in *Schweiz. Arch. Tierheilk.* 85 (1943), pp. 477–81

LAVER, J., *Dress* (John Murray, 1950)

LAVER, J., *Clothes* (Burke, 1952)

LAVER, J., *Costume* (Cassell, 1963)

LEYHAUSEN, P., *Verhaltensstudien an Katzen* (Parey, 1956)

LORENZ, K., 'Der Kumpan in der Umwelt des Vogels', in *J. f. Ornith.* 83 (1935), pp. 137–213, 289–413

LORENZ, K., *On Aggression* (Methuen, 1966)

LYALL-WATSON, M., 'A critical re-examination of food "washing" behaviour in the raccoon', in *Proc. Zool. Soc. London* 141 (1963), pp. 371–94

MONTAGU, M. F. A., *An Introduction to Physical Anthropology* (Thomas, Springfield, 1945)

MORRIS, D., 'Homosexuality in the ten-spined stickleback', in *Behaviour* 4 (1952), pp. 233–61

MORRIS, D., 'The reproductive behaviour of the zebra finch, with special reference to pseudofemale behaviour and displacement activities', in *Behaviour* 6 (1954), pp. 271–322

MORRIS, D., 'The causation of pseudofemale and pseudomale behaviour', in *Behaviour* 8 (1955), pp. 46–57

MORRIS, D., 'The function and causation of courtship ceremonies', in *Fondation Singer Polignac Colloque Internat. Sur L'Instinct, June, 1954* (1956), pp. 261–86

MORRIS, D., 'The feather postures of birds and the problem of the origin of social signals', in *Behaviour* 9 (1956), pp. 75–113

MORRIS, D., *The Biology of Art* (Methuen, 1962)

MORRIS, D., 'Occupational therapy for captive animals', in *Coll. Pap. Lab. Anim. Cent.* 11 (1962), pp. 37–42

MORRIS, D, 'The response of animals to a restricted environment', in *Symp. Zool. Soc. London* 13 (1964), pp. 99–118

MORRIS, D., 'The rigidification of behaviour', in *Phil. Trans. Roy. Soc. London B.* 251 (1966), pp. 327–30

MORRIS, D. (Editor), *Primate Ethology* (Weidenfeld & Nicolson, 1967)

MORRIS, D., *The Naked Ape* (Cape, 1967)

MORRIS, R., and D. MORRIS, *Men and Snakes* (Hutchinson, 1965)

MORRIS, R., and D. MORRIS, *Men and Apes* (Hutchinson, 1966)

MORRIS, R., and D. MORRIS, *Men and Pandas* (Hutchinson, 1966)

OPIE, I., and P. OPIE, *The Lore and Language of Schoolchildren* (Oxford University Press, 1959)

PACKARD, V., *The Status Seekers* (Longmans, 1960)

PICKERING, C., *The Races of Man* (Bohn, 1850)

PIGGOTT, S. (Editor), *The Dawn of Civilization* (Thames and Hudson, 1961)

PIGGOTT, S., *Ancient Europe* (Edinburgh University Press, 1965)

READ, C., *The Origin of Man* (Cambridge University Press, 1925)

RICHARDSON, L. F., *Statistics of Deadly Quarrels* (Stevens, 1960)

RUSSELL, C., and W. M. S. RUSSELL, *Human Behaviour* (André Deutsch, 1961)

SCHUTZ, F., 'Homosexualität und Prägung', *Psychol. Forschung* 28 (1965), pp. 439–63

SCOTT, J. P., 'Critical periods in the development of social behaviour in puppies', *Psychosom. Med.* 20 (1958), pp. 45–54

SCOTT, J. P., and J. L. FULLER, *Genetics and the Social Behaviour of the Dog* (Chicago University Press, 1965)

SEGAL, R., *The Race War* (Cape, 1966)

SLUCKIN, W., *Imprinting and Early Learning* (Aldine, 1965)

SMAILES, A. E., *The Geography of Towns* (Hutchinson, 1953)

SMITH, A., *The Body* (Allen & Unwin, 1968)

STENGEL, E., *Suicide and Attempted Suicide* (Penguin, 1964)

STORR, A., *Human Aggression* (Penguin Press, 1968)

TAX, S. (Editor), *The Evolution of Man* (Chicago University Press, 1960)

TINBERGEN, N., *The Study of Instinct* (Oxford University Press, 1951)

TINBERGEN, N., *The Herring Gull's World* (Collins, 1953)

TURNER, E. S., *All Heaven in a Rage* (Michael Joseph, 1964)

WASHBURN, S. L. (Editor), *Social Life of Early Man* (Methuen, 1962)

WASHBURN, S. L., and I. DEVORE, 'Social behaviour of baboons and early man', in *Social Life of Early Man* (Editor: S. L. Washburn), (Methuen, 1962)

WEST, D. J., *Homosexuality* (Aldine, 1968)

WHITMAN, C. O., *The Behaviour of Pigeons* (Carnegie Institution, 1919)

WICKLER, W., 'Socio-sexual signals and their intra-specific imitation among primates', in *Primate Ethology* (Editor: D. Morris), (Weidenfeld & Nicolson, 1967)

WICKLER, W., *Mimicry in Plants and Animals* (World University Library, 1968)

WODDIS, G. M., 'Depression and crime', in *Brit. J. Delinquency* (1957), pp. 85–94

ZEUNER, F. E., *A History of Domesticated Animals* (Hutchinson, 1963)

ZUCKERMAN, S., *The Social Life of Monkeys and Apes* (Kegan Paul, 1932)